CARPENTRY & WOODWORKING

CARPENTRY & WOODWORKING

Dick Demske

CREATIVE HOMEOWNER PRESS®   24 PARK WAY, UPPER SADDLE RIVER, NEW JERSEY 07458

Manufactured in United States of America

Current printing (last digit)
10 9 8 7 6 5 4

Produced by Roundtable Press, Inc.

Editorial: Judson Mead
Design: Betty Binns Graphics
Jacket photo: David Arky
Tools on cover courtesy the Garrett Wade Co. Inc.

LC: 83-15094
ISBN: 0-932944-62-0 (paper)
        0-932944-63-9 (hardcover)

CREATIVE HOMEOWNER PRESS®
BOOK SERIES
A DIVISION OF FEDERAL
MARKETING CORPORATION
24 PARK WAY,
UPPER SADDLE RIVER, NJ 07458

INTRODUCTION

After countless thousands of years of use, wood remains a favorite building material for professionals and amateurs alike. It is readily available and relatively inexpensive (although some more exotic species can be as costly as they are beautiful). It is easily cut, shaped, fitted, and joined without high-priced machinery. Depending on the final use, wood can be left unfinished, painted, stained, or clear-finished to highlight the rich warmth of the natural wood.

Engineered wood-based panel products, such as plywood, hardboard, and particleboard, add another dimension to the forests' products. They simplify and speed up building construction, while adding considerable rigidity to a structure. They also simplify and strengthen the construction of furniture, shelving, built-ins, and countless other projects.

In a broad sense, carpentry and woodworking both describe working with wood. But the terms have come to take on slightly different connotations. A carpenter is a woodworker who builds or repairs wood structures. Carpentry is further divided into rough and finish carpentry. Rough carpentry includes framing, sheathing, decking, and other operations in the structural and exterior parts of a house or other building. Finish carpentry refers to those operations that enhance the interior appearance of the building, such as paneling, cabinetry, and trim.

Woodworking in its narrower meaning is perhaps better termed cabinetmaking. It is an extension of carpentry, and encompasses the making of furniture, cabinets, and countless other fine wood products. This type of woodworking generally requires more skill, patience, and precision. It also calls for more knowledge of woods, wood joints, wood forming, and wood finishing than does simple carpentry.

Whether rough carpentry or fine cabinetmaking, there are no profound mysteries in working with wood. Can you master it? Very likely. What you need is some basic knowledge of wood itself—the various types, species, grades, and variations available—as well as the many other wood products that are available to the do-it-yourselfer, plus a knowledge of the tools and techniques involved.

The first four chapters of this book provide you with the fundamental knowledge of wood, wood products, and the common tools for working them. The subsequent chapters detail the techniques of basic carpentry and woodworking, including basic joinery; the fasteners and adhesives used to hold wood projects (from a frame house to a simple shelf) together; methods of sanding wood to get it ready for finishing; and the various finishes available to give a project the look you want. The concluding section of the book puts these techniques to work with simple projects you can build to try out new-found skills.

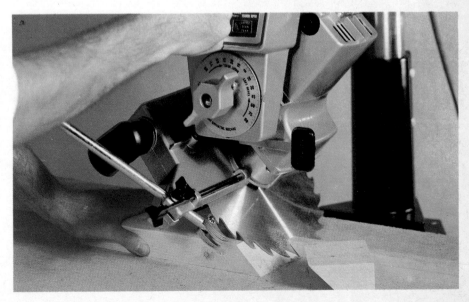

Contents

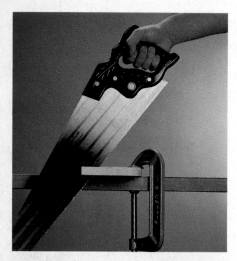

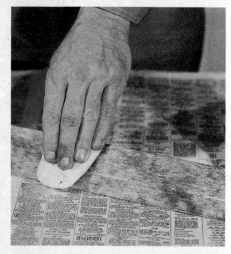

WOODWORKING PROJECTS

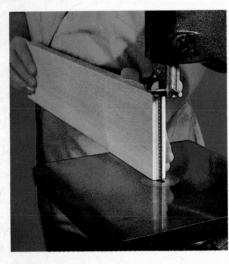

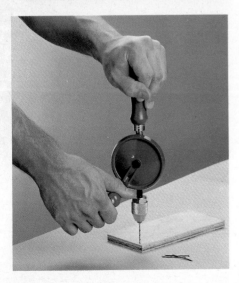

1 WOOD

Our forests are a truly remarkable natural resource. Covering almost one-third of the earth's land surface, they serve an important role in conserving soil and water and providing a refuge for wildlife. And they provide a unique recreational refuge for an increasingly urbanized world. Forests also supply a basic raw material —wood—which from earliest times has furnished mankind with necessities of life as well as countless conveniences and comforts.

If you were to look at a piece of wood under a microscope, you would see that it is made up of thousands of hollow cells. These natural building blocks are formed from tiny cellulose fibers, about three million per cubic inch. These tough fibers and the cells they form are cemented together by a natural glue, *lignin*. Lignin bonds so tenaciously to cellulose that the two were once considered a single substance. No other building material is structured this way, and no other building material possesses qualities quite like those of wood.

Wood displays incredible strength. A small block of wood, say $1 \times 1 \times 2\frac{1}{4}$ inches, can support 10,000 pounds —more than the weight of five compact cars. Wood is actually stronger, pound for pound, than steel, deriving its strength from the natural strength of its cells. And the lignin that cements the cells together is not only strong, it is also elastic. For this reason, wood can, up to a point, bend without breaking.

Again because of its cellular structure, wood is an excellent natural insulator. It would take a concrete wall 5 feet thick to equal the insulating quality of just 4 inches of wood. Wood insulates 6 times better than brick, 15 times better than concrete, and an amazing 1,770 times better than aluminum. Those wood cells contain millions of tiny air spaces, and air is one of the best insulators known.

Wood is also a valuable acoustical material. It can reflect or absorb sound waves and is often used in public buildings for sound control. Many fine musical instruments are made of wood, because wood can be crafted to resonate at frequencies common to music; this is due in part to the unique cellular structure.

The lignin that holds wood cells together is largely impervious to water and unaffected by extremes of heat and cold. Because of this wood can last for centuries—even millennia —when properly used. The Old Ironworks House in Saugus, Massachusetts, still stands as majestic as ever despite the lashings of almost 350 New England winters. The 1660 saltbox in East Hampton, Long Island, exudes the same mellow warmth today as it did when it inspired John Howard Payne to compose "Home Sweet Home." In Venice, Italy, wooden piles found recently were intact after having been under the city's streets for 1,000 years. And timbers more than 2,700 years old have been discovered in the tomb of King Gordius near Ankara, Turkey.

Add to these qualities wood's beauty, versatility, workability, and ready availability, and it becomes obvious that this is indeed nature's miracle material.

The structure of wood

A tree in cross-section has three well-defined features in succession from the outside to the center: bark, wood, and pith. *Bark* is divided between the outer, corky, dead part (which varies greatly in thickness with species and age of trees) and a thin, inner, living part. The *wood*, in most species, is clearly differentiated into sapwood and heartwood. The *pith* is a small central core, darker in color, which represents primary growth formed when stems or branches elongate.

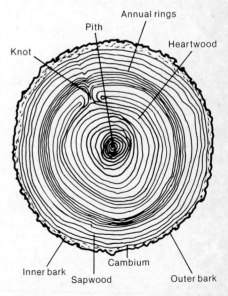

The cross section of a log shows the principal parts of live wood.

Pith is not separated from the wood in rough lumber and timbers, but it is excluded from finish lumber.

Most branches originate at the pith, and their bases are intergrown with the wood of the trunk as long as they are alive. These living branch bases constitute intergrown knots. After the branches die, their bases continue to be surrounded by the wood of the growing trunk, and these constitute loose or encased knots. After the dead branches fall off the tree, the dead stubs become overgrown, and clear wood is formed.

Between the bark and the wood is a thin layer of living cells, visible only under a microscope, called the cambium, in which all growth in thickness of the bark and the wood takes place by cell division. As the diameter of the woody trunk increases, the bark is pushed outward. With many species, there is sufficient difference between the wood formed early in a growing season and that formed late to produce obvious, well-marked annual growth rings. The age of a tree may be determined by counting these rings at any cross-section of the trunk. If the growth of a tree is interrupted—by drought, defoliation by insects, or some other cause—more than one ring may be formed in the same season. In that case, the inner rings usually do not show sharply defined boundaries and are termed "false rings."

The inner part of the growth ring, formed early in the growing season, is called *springwood*; the outer part, formed later in the season, is *summerwood*. Springwood is characterized by cells with relatively large cavities and thin walls. Summerwood cells have smaller cavities and thicker walls. In wood species where growth rings are prominent, springwood differs markedly from summerwood in physical properties. Springwood is lighter in weight, softer, and weaker than summerwood; it shrinks less across the grain and more along the grain.

Sapwood, located next to the cambium, contains living cells and plays an active role in the life processes of the tree. The sapwood layer varies in thickness and in the number of growth rings contained in it. As a rule, vigorously growing trees have wide sapwood layers.

Heartwood consists of inactive cells formed by changes in the living cells of the inner sapwood rings. The cavities of heartwood sometimes

Working Characteristics and Strength Properties of Hardwoods

Species	Hardness	Weight (dry)	Freedom from shrinkage, swelling	Freedom from warping	Ease of working	Nail holding	Decay resistance of heartwood	Bending strength	Stiffness	Toughness	Common Uses
ash:											
black	B	B	C	B	C	A	C	B	B	A	implements, cooperage, containers, furniture
white	A	A	B	B	C	A	C	A	A	A	implements, containers, furniture, veneer
aspen	C	C	B	B	A	C	C	C	B	C	boxes, lumber, pulp, veneer
basswood	C	C	C	B	A	C	C	C	B	C	woodenware, boxes, veneer, lumber
beech	A	A	C	C	C	A	C	A	A	A	flooring, furniture, woodenware, veneer
birch	B	A	C	B	C	A	C	A	A	A	flooring, furniture, millwork, veneer
cherry	B	B	B	A	B	A	C	A	A	B	furniture, woodenware, paneling, gunstocks
cottonwood	C	C	B	C	B	C	C	C	B	C	pulpwood, containers, woodenware, lumber, veneer
elm:											
rock	A	A	B	B	C	A	C	A	A	A	furniture, containers, veneer, cooperage
slippery	B	B	B	C	C	A	C	B	B	A	containers, furniture, veneer
hackberry	B	B	C	B	C	A	C	B	C	A	furniture, veneer, containers
hickory	A	A	C	B	C	A	C	A	A	A	handles, athletic goods, implements, flooring
locust	A	A	A	B	C	A	A	A	A	A	poles, posts, ties, containers, fuel
magnolia	B	B	B	B	B	A	C	B	B	B	furniture, veneer, containers, millwork
maple	B	B	B	B	C	A	C	B	B	B	furniture, woodenware, pulpwood, fuel
oak:											
red	A	A	B	B	C	A	C	A	A	A	flooring, furniture, veneer, posts, millwork
white	A	A	C	B	C	A	A	A	A	A	furniture, cooperage, millwork, veneer, flooring
sycamore	B	B	B	C	C	A	C	B	B	B	furniture, veneer, cooperage, containers
tupelo	B	B	C	C	C	A	C	B	B	B	containers, furniture, veneer, cooperage
walnut	B	A	B	A	B	A	A	A	A	A	furniture, gunstocks, interior finish, veneer

Key A: relatively high in the quality   **B:** intermediate in the quality   **C:** relatively low in the quality   (These letters do not refer to lumber grades.)

contain deposits of various materials that give heartwood a much darker color than sapwood. These materials often plug the cells, and lumber cut from such heartwood is more durable when used in exposed situations than that from sapwood. Sapwood should always be treated with a preservative when exposed to weather.

Hardwoods and softwoods

Native species of trees fall into one of two general classifications: hardwood and softwood. *Hardwoods* have broad leaves and (except in very warm regions) are deciduous, shedding their leaves at the end of each growing season. *Softwoods* have scalelike leaves (as do cedars) or needles (as do pines) and—except for cypress, larch, and tamarack—

are evergreen, the leaves or needles remain on the branches all year. Softwoods are known also as conifers, because all native species bear cones of one kind or another.

The terms "hardwood" and "softwood" are often misunderstood; they have no direct application to the hardness or softness of the wood. Actually, the wood of such hardwood trees as cottonwood and aspen is softer than white pine and fir. On the other hand, such softwoods as longleaf pine and Douglas fir produce wood that is as hard as that of the hardwoods yellow poplar and basswood.

The principal difference between the two classifications is in the density of the wood. The millions of tiny cells that make up wood are packed more tightly together in hardwoods;

also, the cell walls are thicker, making the wood denser, heavier, and, generally, stronger. Softwoods, with thinner cell walls, are lighter and more porous.

Softwoods are used principally for construction. Among the more common of these are northern white pine, Idaho white pine, white fir, ponderosa pine, sugar pine, redwood, and cedar—all of which are easily worked. The heavier Douglas fir and southern yellow pine are frequently used for joists and rafters of greater length than can be spanned by equal-sized members of the other softwoods. Hardwoods are widely used for interior trim and flooring, as well as for most wood furniture. Traditional favorites are ash, oak, maple, mahogany, cherry, birch, elm, and walnut.

Working Characteristics and Strength Properties of Softwoods

Species	Hardness	Weight (dry)	Freedom from shrinkage, swelling	Freedom from warping	Ease of working	Nail holding	Decay resistance of heartwood	Bending strength	Stiffness	Toughness	Common Uses
cedar:											
eastern red	B	B	A	A	B	B	A	B	C	B	posts, paneling, wardrobes, chests
northern white	C	C	A	A	A	C	A	C	C	C	poles, posts, tanks, woodenware
southern white	C	C	A	A	A	C	A	C	C	C	posts, poles, boats, tanks, shingles
western red	C	C	A	A	C	A	C	C	C	shingles, siding, poles, boats, paneling, millwork
cypress	B	B	A	B	B	B	A	B	B	C	millwork, siding, tanks, poles, shakes
Douglas fir	B	B	B	B	B	A	B	A	A	B	construction, plywood, millwork, flooring, piling, poles
fir: balsam	C	C	B	B	B	C	C	C	C	C	light construction, pulpwood
white	C	C	A	B	B	C	C	B	A	C	light construction, containers, millwork
hemlock:											
eastern	B	C	A	B	B	B	C	B	B	C	construction, containers, pulpwood
western	B	C	B	B	B	B	C	B	A	B	construction, pulpwood, containers, flooring
larch	B	A	B	B	C	A	B	A	A	B	construction, poles, ties, millwork
pine:											
Idaho white	C	C	B	A	A	C	C	B	B	C	millwork, construction, siding, paneling
lodgepole	C	C	B	C	B	B	C	B	B	C	poles, lumber, ties, mine timbers
northern white	C	C	A	A	A	C	B	C	C	C	millwork, furniture, containers, paneling, siding
ponderosa	C	C	A	A	A	B	C	C	C	C	millwork, construction, veneer, paneling
southern yellow	B	A	B	B	B	A	B	A	A	B	construction, siding, cooperage, ties, plywood
sugar	C	C	A	A	A	C	C	C	C	C	millwork, construction, containers, siding
redwood	B	C	A	A	B	B	A	B	B	C	siding, tanks, millwork, outdoor furniture, decks
spruce:											
eastern	C	C	B	A	B	B	C	B	B	C	construction, pulpwood, millwork, containers
Engelmann	C	C	A	A	B	C	C	C	C	C	light construction, poles, pulpwood
Sitka	C	C	B	A	B	B	C	B	A	B	construction, millwork, containers, pulpwood

Key A: relatively high in the quality   **B:** intermediate in the quality   **C:** relatively low in the quality   (These letters do not refer to lumber grades.)

The accompanying tables show the characteristics of various hardwoods and softwoods.

From forest to lumber

When trees are harvested, they are cut into large logs and transported to a sawmill. Bark is removed by mechanical or hydraulic means, then a log is firmly fastened on a carriage with sufficient edge protruding to accommodate the first cut. A skilled sawyer stationed at a control panel where he can view the log clearly then decides how the log should be cut into the best possible assortment of lumber grades, thicknesses, and widths. The movable carriage, at the operator's command, thrusts the log forward through a stationary bandsaw, then returns to begin a subsequent cut. The first cut produces a slab, generally reduced to chips for making pulp and paper. The second cut, usually 1 inch thick, reveals more clearly the interior quality of the log as well as the grades of lumber to be expected from the next cut. After five or six cuts, the sawyer rotates the log in the carriage to cut into a face adjacent or opposite to the first cuts. The sawing and rotating are repeated until the best pieces are removed and the log has been reduced to a heavy square or rectangular piece. This may be marketed as is, or it may be cut into smaller pieces by edgermen or resawyers operating other saws in the mill.

When lumber is cut tangent to the annual rings of a log, it is called *plainsawed* (softwoods so cut are also known as *flat-* or *slash-grained*). Lumber cut radially to the rings is *quartersawed* (*edge-* or *vertical-grained* in softwoods). In actual practice, lumber with rings at 45° to 90° with the surface is considered quartersawed, and lumber with rings at angles of 0° to 45° with the surface is plainsawed. Hardwood lumber in which the annual rings make angles of 30° to 60° with the faces is also known as *bastard-sawn*.

Rift sawing is similar to quartersawing except that long and short boards are cut alternately from the log. The angle of the cuts vary slightly from one board to another,

and small wedges of wood are wasted between the boards. However, this waste is less than in quartersawing.

Either of these cuts produces a strong, physically attractive board. The inherent waste does add significantly to the cost.

While either plainsawed or quartersawed lumber is satisfactory for many purposes, each has advantages. Plainsawed lumber is generally cheaper because it is easier to cut from the log and produces less waste. Figure patterns resulting from the annual rings are more conspicuous. Knots that may occur affect the surface appearance less than in quartersawed boards, and the knots weaken the board less. However, there are generally more knots in plainsawed than quartersawed lumber. Plainsawed lumber also shrinks and swells less in thickness.

Quartersawed (and rift sawed) lumber shrinks and swells less in width. It also twists and cups less, and checks and splits less in seasoning and in use. Quartersawed lumber wears more evenly and, in some species, is impervious to liquid. Raised grain caused by separation in the annual rings is less pronounced than in plainsawed, while figures caused by interlocked grain and wavy grain are brought out more conspicuously, providing interesting surface patterns.

Seasoning

The moisture content of lumber coming from the sawmill varies according to the wood species, but it is generally too high to be suitable for building. Wood shrinks as it dries and swells as it absorbs moisture. Much of this change in dimension can be avoided by drying, or *seasoning*, wood to a suitable moisture content.

Lumber may be seasoned by natural air-drying, by controlled air-drying (kiln-drying), or by the use of various chemicals (among them salt and urea) in conjunction with one of the other two methods. Factors determining the method of seasoning include the time available for drying,

Bark is peeled from logs with a jet of water under high pressure.

Quartersawn (bottom) and plainsawn (top) boards cut from a log.

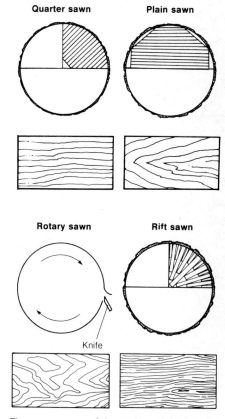

The appearance of the grain and the relative strength of boards is determined by the wood itself, and by the way the boards are sawn. Rotary-sawn wood is used to make plywood.

the wood species, and the intended use of the wood. Kiln-drying is the common commercial practice. Chemical seasoning in combination with air- or kiln-drying is used for certain high-quality lumber.

Kiln-dried lumber is generally preferred for furniture projects because it offers more protection against shrinkage (which can cause joints to open). Air-dried lumber is often the choice for projects that will be exposed to the weather.

Lumber sizes

Mastering lumber sizes is not difficult, but it does take a little getting used to since the numbers do not always mean what they say. A 2×4, for example, actually measures 1½ × 3½ inches, while a 1 × 10 is ¾ inch thick and 9¼ inches wide. A ½ × 6 board is indeed ½ inch thick, but it is only 5½ inches wide.

Your lumberyard is not violating any truth-in-advertising statutes. That 2×4 was a full 2 inches by 4 inches when it was cut from the log. It slimmed down during the seasoning process, then it was surfaced (*dressed*) on both sides and both edges to make it uniform in size and ready to use. The sizes for seasoned lumber, uniform throughout the United States, are given in the accompanying table.

For boards less than nominal 1-inch thickness, the true thickness is given, but the nominal width is still used. Another fairly common thickness is ⁵/₄ (five quarter). This is a board that is cut to an actual 1¼ inches thick and finish dressed to 1¹/₁₆ inches. Use ⁵/₄ wood when you need a board that is 1 inch thick; you just plane down the difference.

Dressed lumber sizes were the subject of some confusion and controversy in the past until uniform standards were established by the industry with the assistance of government agencies. It seems likely that the controversy will be rekindled when lumber goes metric, and a 2 × 4 becomes a nominal 5 × 10 (centimeters)—when dressed, 38 × 89 millimeters.

Grading

Lumber quality varies, both between and within the various wood species. For this reason, grade standards have been established by a number of lumber manufacturers' associations in accordance with government guidelines. Softwood grading is based on a visual inspection and is primarily a judgment of end-use strength and, to a lesser extent, appearance. The official grading association stamp on a piece of lumber is an assurance of its assigned grade. The stamp may also contain information on the wood species, the moisture content, and the mill where it was cut.

Grading terminology may differ depending on the certifying agency and, in some cases, the wood species. Generally, there are two basic classifications: select and common. In softwoods, select grades include:

- **B & Better** (or **No. 1** and **No. 2 clear**), with only minute imperfections

- **C select**, with small knots or other blemishes

- **D select**, with slightly larger imperfections (which can be concealed by paint)

Common grades are most frequently designated by number—the lower the number, the higher the quality. Other grade designations include *Select Structural, Construc-*

Lumber Sizes

Nominal size	Actual dry size (in inches)
1×2	¾ × 1½
1×3	¾ × 2½
1×4	¾ × 3½
1×6	¾ × 5½
1×8	¾ × 7¼
1×10	¾ × 9¼
1×12	¾ × 11¼
2×2	1½ × 1½
2×3	1½ × 2½
2×4	1½ × 3½
2×6	1½ × 5½
2×8	1½ × 7¼
2×10	1½ × 9¼
2×12	1½ × 11¼
3×3	2½ × 2½
3×6	2½ × 5½
4×4	3½ × 3½
4×6	3½ × 5½

Knots: a common characteristic.

More knots through multiple faces.

Wane: bark or absence of wood.

Splits and checks are cracks.

Shakes are lengthwise separations.

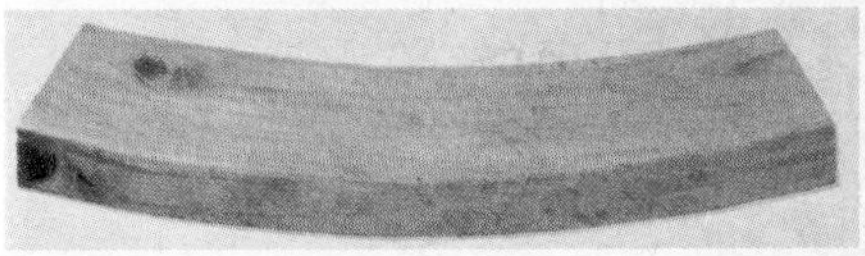

Bow: a curve in the wide face.

Crook: a curve in the narrow face.

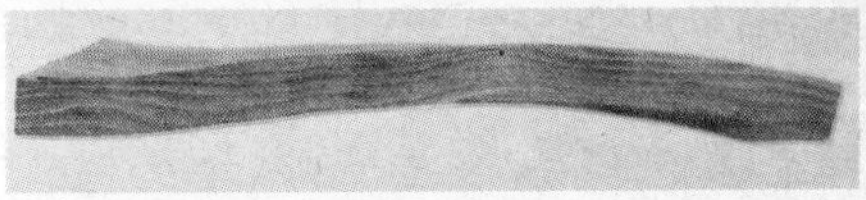

Twist: a spiral of all four faces.

White speck caused by fungus.

Dote: also called rot or unsound wood.

Characteristics of typical lumber grades.

Select Structural: limited knots.

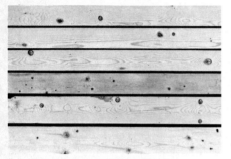

No. 1: slightly larger encased knots.

No. 2: well-spaced knots, some holes.

No. 3: well-spaced knots, more holes.

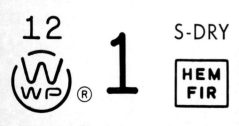

Grade stamp shows grade, species, moisture content (S-DRY means 19% or less), mill (12), and grader (WWP).

Characteristics of redwood lumber grades.

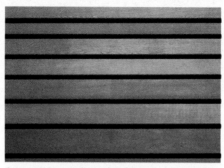

Clear All Heart: free of defects.

Clear: contains some sapwood.

Construction Heart: allows some knots.

Construction Common: knots and sapwood.

Merchantable Grade: sapwood, large knots.

tion, *Standard*, *Utility*, and *Stud*. These are determined primarily by the number, spacing, and tightness of knots in a piece of lumber.

Hardwood grading takes into account the yield and size of cuttings with one clear face that can be cut from the lumber. The highest grades are *firsts* and *seconds*. Hardwoods for construction are grouped into three general classes: *Finish*, *Construction and Utility Boards*, and *Dimension*. Finish is subdivided: *A Finish* has one face practically clear, while *B Finish* may contain small surface checks, streaks, and other minor blemishes. Construction and Utility Boards are graded *No. 1*, *No. 2*, or *No. 3* based on the amount of wane, checks, knots, and other imperfections. Dimension grades (nominally 2 inches or more thick) are either *No. 1* or *No. 2*, depending on the number of imperfections.

If all these terms and numbers have set your head spinning, relax! As a do-it-yourself carpenter or woodworker, you can follow a few simple rules when selecting lumber for your projects. Use the highest grades when good appearance or extra strength is the foremost requirement. For most jobs, the middle grades will probably be the choice. The poorest grades are acceptable when economy, rather than strength or appearance, is a primary factor.

Buying lumber

Before you go to the lumberyard, make sure you know exactly what you need. Plan and measure your project carefully, and determine what grade will best suit the strength and appearance required. (If in doubt on this point, the lumberyard salesman should be able to help.) You may have to order or shop around if you need a certain wood species for a project.

If possible, select each piece of lumber yourself; sight along the edges and reject those that are badly bowed, crooked, or twisted or that show other serious imperfections. When ordering a quantity of lumber, specify the lengths wanted. If you need fifteen 8-foot 2×4s, for example, say just that—not 120 feet of 2×4. In the latter case, the lumberyard might ship you ten 12-foot 2×4s, leaving you with a lot of waste and the need to order more material.

Many lumberyards will cut lumber to specified lengths, often for a minimal charge. But don't expect them to do this on a busy Saturday afternoon. Plan ahead, and let them know exactly what you want. Your measurements are critical, so take your time and be accurate.

Lumber is sometimes sold by the linear or running foot, but the more common standard for determining cost is the board foot. A *board foot* is a unit nominally 1 inch thick × 1 foot wide × 1 foot long. To calculate board feet, multiply the nominal thickness (inches) by the nominal width (inches) by the actual length (feet) and divide by 12. Thus, an 8-foot long 1×12 contains 8 board feet ($\frac{1 \times 12 \times 8}{12} = 8$). So does a 12-foot 2×4 ($\frac{2 \times 4 \times 12}{12} = 8$), and an 8-foot 2×6 ($\frac{2 \times 6 \times 8}{12} = 8$). If you order 2×4s totaling 120 feet, you will be charged for 80 board feet of lumber ($\frac{2 \times 4 \times 120}{12} = 80$). If your order is for 96 feet of 1×10s, you will also be charged for 80 board feet ($\frac{1 \times 10 \times 96}{12} = 80$). If both orders are for the same grade and species, the board foot price should be the same.

Buying fine woods

Unlike construction-grade wood that is cut to size and then allowed to dry and shrink below its nominal size, fine woods meant for furniture or cabinetmaking are cut larger than nominal size. This is true of all hardwoods and of furniture-grade softwoods such as pine.

Furniture-grade wood is kiln- or air-dried after sawing. The drying process reduces the bulk of the wood by approximately 10 percent. The boards are then sold in designations called *quarter stock*.

It should be noted that you probably will not have access to all of the types of wood discussed below. Many are in limited supply and available only on special order or from suppliers that specialize in furniture-grade stock. Normally, a lumberyard stocks wood that is indigenous to the immediate area. One or another type of pine is usually plentiful, and you may find that a yard will stock at least one form of mahogany. Because of the increased costs, dealers usually will not handle wood that has to be shipped a great distance. If you are willing to accept shipping costs as part of the price, you will usually be able to order any wood you want. However, you will have to accept the boards that arrive as is, even if they are not exactly what you had envisioned.

Types of wood

One of the obvious characteristics of wood is *grain*. Although most people respond to the appearance of the grain, for the wood finisher, the primary concern is whether the grain is open or closed. Some wood grain looks fine but has large pores; this creates an open grain. Small pores in a wood create a closed grain.

In furniture making, three of the most commonly used open-grained woods are walnut, oak, and mahogany.

To create the smooth, even surface usually desired in a furniture finish, open-grained wood is commonly filled before finishing. A soft filler material is rubbed on and into the wood to fill the pores. If the filler is not used, the finish applied will seep into the open pore areas in greater quantities than will soak into the closed areas. This will create an unsightly, rippled effect. The unfilled surface also will be rough; filler makes the surface smooth and even.

Popular closed-grained furniture woods are birch, maple, pine, fir, beech, gum, and poplar. These woods usually are not filled or are filled with only a very thin filler. Examples of closed-grained wood with very tiny pores are basswood and holly. Basswood is seldom used for furniture, but holly is sometimes used in cabinets.

The following are the types of wood you are most likely to use in your projects—and a few that are more exotic.

Ash. Ash is an extremely hard hardwood. Because it is so hard, the wood is usually used for baseball bats and tool handles. It is creamy

End grain (left), grain from quarter sawn wood (center), plain sawn (right).

Black Cherry

Western Red Cedar

White Ash

Yellow Birch

Black Walnut

Redwood

Douglas Fir

Ponderosa Pine

Sugar Maple

White Oak

white in color, with very little grain pattern. It is, however, open-grained and must be filled before finishing. Although strong, it can be worked quite easily. It also is remarkably flexible; you can bend it more than other woods with little danger of its splitting.

Basswood. This is a white wood that normally has very little grain showing. It is a light, easy-to-work wood. It is not as hard as some other hardwoods, but it is closed-grained. Basswood is normally used in making the interior parts of cabinets and as a core material, or *banding*, for plywood. It is not normally used as a finish material, but it does take a good finish.

Birch. This attractive hardwood is normally yellow-white to red. The commercially cut wood is yellow birch. It is closed-grained and does not need to be filled. A strong wood, it is still fairly easy to work. However, it is often more expensive than mahogany.

Cedar (western red). This wood is used mostly outside the house for shingles and small structures such as storage buildings; it has a very high resistance to weather and rot. It ranges in color from pink-brown to brown and is easy to work and finish. Some people use cedar shingles inside the house as a decorative element on one or more of the walls. No filler is required for the wood.

Cherry (American black). This wood can vary in color from light to reddish brown. Cherry has a fine grain pattern, is closed-grained, and accepts clear finishes, such as varnish, very well. It is easy to work with and is a popular wood for furniture making.

Cypress. Another wood used primarily outdoors, where particular resistance to decay is desired, is cypress. It is available in shades from brown to almost black and with red overtones. The wood can feel oily to the touch and, while easy to work, it is not the kind of wood one nor-

mally considers for items inside the house.

Fir. This wood is basically a tan or yellow with darker brown streaks. Its grain pattern is very prominent. It is a strong wood that is not that easy to work with. Because of its grain pattern, it is often covered with opaque finishes.

Gum. Used more commonly in making veneer than as a building material for projects, gum is a reddish brown, attractive wood. While it is not particularly strong, it can be used with success to build cabinets. Availability of the wood is limited. Gum is closed-grained—no filling is required.

Hickory. A wood that is very hard and difficult to work with, hickory is surprisingly vulnerable to decay and insect infestation. It is seldom used for furniture or for household woodwork, but it is commonly used in tool handles.

Mahogany (African and Honduras). This is considered one of the finest furniture woods. These days, fine woods are very expensive and mahogany will carry a premium price.

Although mahogany is usually thought of as a reddish wood, you will find boards that range from a golden brown (Honduras) to a red brown (African). Mahogany boards are most commonly available in golden brown. Although this wood is of medium hardness, it is very strong and is often used for furniture because of its attractive color and fine grain pattern. It accepts all types of finishes well, but it is open-grained, so it must be filled before the finish can be applied.

Maple (true, hard). Available in a variety of colors ranging from nearly white to red-brown, maple is available in various configurations. While maple is normally straight-grained, it is also available in curly, burl, and "fiddleback" patterns, among others. Bird's-eye maple, for example, has a small swirling pattern and is used for interior paneling in the Rolls Royce. Maple is very tough, strong,

hard, and closed-grained—it takes finishes very well.

Oak. There are two types of oak— red oak and white oak. As the name implies, red oak is red (actually a reddish tan or brown). White oak is lighter, usually a grayish tan.

Oak is a very hard hardwood, and it is well known for its durability. Oak is used in furniture and flooring because it is attractive and will stand up to long-term, heavy use and wear. However, it does have very large pores that must be well filled, otherwise the wood will not accept a stain or other finish well.

Because oak is so hard, it is difficult to work with. If you have ever attempted to drive a light nail into a piece of oak, this fact has been made abundantly clear. However, with the right tools and equipment you can work with it.

Pine. This softwood is one of the most popular furniture woods because it is plentiful, relatively cheap, and easily finished. It is often a white or cream color, but the color can vary considerably from region to region and from species to species.

Pine comes in various grades, ranging from Clear, which is the best, through Select No. 1, 2, and 3, to Common pine. The Clear grades have no knots or blemishes; Select comes with a few; Common has the most. It is heartily suggested that you pick through the stack of boards at the lumberyard so that you can get the best of what is available. For example, under the Common grade heading you can find pieces that

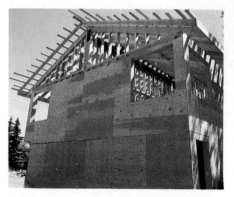

Plywood sheathing allows builder to cover large areas in little time.

have varying degrees of blemishes, and you may find exactly what you need without having to buy the better grades.

Pine is normally available in an abundance of sizes. Furniture grades are available in quarter stock in many thicknesses and widths. Construction grades from Clear to Common are available in 1 × 2, 1 × 3, and 1 × 4, then up to 1 × 18 in 1-inch increments. It is also available in standard 2x stock in all sizes and 4x stock in several sizes. The greatest advantage of this wood is, quite simply, that it is readily available in so many grades and sizes.

Redwood. This expensive, reddish brown stock is a big favorite for outdoor projects. It can be finished, but if left unfinished it will weather to a silvery gray. It is light and strong, has outstanding resistance to decay, is quite soft and easy to work with, and is used for sheet or plank paneling, woodwork, shelving, and furniture.

Walnut (American black). A popular furniture wood, walnut is usually a chocolate brown color. The grain is prominent and open. Walnut is of medium hardness. Much of the commercial production is used for veneer; boards are scarce and expensive.

Wood preservatives

Some wood species—notably cedar and redwood—are extremely rot-resistant; but, unless they are treated with preservative, most other woods will decay quickly if they are in contact with the ground or left exposed to extreme weather conditions. The most effective treatment is done by a pressure process at the mill or some other commercial firm. There are three broad classes of preservatives used in pressure-treating:

- **waterborne solutions:** clean, odorless, and paintable but, since the preservative is carried by water, the wood must be redried after treatment, either by natural air-drying or kiln-drying
- **creosote** and **mixtures of creosote and coal tar:** used for wood in ex-

treme exposure conditions; creosoted wood is dark brown to black in color, and it has a rather unpleasant odor
- **pentachlorophenol:** used to protect wood exposed to harsh climatic conditions; will turn wood brown in color, but eventually takes on a silver tone; some treatments utilizing light solvents provide surfaces that are clean, dry, and paintable or stainable

Superficial application of preservatives by such methods as dipping, brushing, or spraying is generally ineffective, since the preservative does not penetrate the wood deeply enough to afford adequate protection against decay. If pressure-treated lumber is unavailable, the next best treatment is to immerse the lumber to be used in creosote or pentachlorophenol (available at many hardware and building supply stores) and let it soak for two to three weeks.

Water-repellent substances are marketed under several trade names and can be found at any hardware or paint store. While these are not really preservatives, they are helpful in protecting wood from exposure to the elements. Follow manufacturer's directions for their application.

Plywood

Plywood has long been a favorite material of do-it-yourselfers. It is rigid and strong, and it weighs far less than most metals, solid lumber, or hardboard of equal strength. It does not split, chip, crumble, or crack through and has high impact resistance. It comes dry from the mill, never green. Even when totally saturated with moisture, a plywood panel swells less than two tenths of a percent across or along the grain—resisting expansion and contraction under most unfavorable conditions.

Plywood is manufactured by laminating sheets of veneer together, with the grain of adjacent sheets at right angles. The adhesive bond between plies is formed under pressure, making the bond at least as strong as the wood itself. Panels are

Razor-sharp knives cut plys from spinning log to make plywood.

constructed of an odd number of layers, so that both face plies have grain running in the same direction. Each layer may consist of a single ply or two or more plies laminated with grain directions parallel. The cross-lamination distributes wood's high along-the-grain strength in both directions, creating a panel that is virtually split-proof and puncture-proof.

Plywood is widely used for such construction purposes as wall and roof sheathing, siding, flooring, and interior paneling. Because it comes in large panels (4 × 8 feet is the most common, but larger sizes are also available), it enables the user to cover large areas quickly, saving time and labor. It is also the first choice for many furniture projects, built-ins, shelving, and countless other home workshop applications. The exact panel sizes mean that waste can be kept to a minimum, and the material is easily worked with common carpentry tools.

Plywood is made from both hardwoods and softwoods.

Softwood plywood

Softwood plywood is made in two types: Exterior, bonded with waterproof glue, and Interior, bonded with moisture-resistant glue. *Exterior* should be used when the panel will be exposed to the weather. *Interior* will not be significantly affected by occasional wettings during construction, but it should not be used where it will be permanently ex-

Classification of Species for Softwood Plywood

Group 1	Group 2	Group 3	Group 4	Group 5
apitong (a), (b)	cedar, Port Orford	maple, black	alder, red	aspen
beech,	cypress	mengkulang (a)	birch, paper	bigtooth
American	Douglas fir 2 (c)	meranti, red (a), (d)	cedar, Alaska	quaking
birch	fir	mersawa (a)	fir, subalpine	cativo
sweet	California red	pine	hemlock, eastern	cedar
yellow	grand	pond	maple, bigleaf	incense
Douglas fir 1 (c)	noble	red	pine	western red
kapur	Pacific silver	Virginia	jack	cottonwood
keruing (a), (b)	white	western white	lodgepole	black (western poplar)
larch, western	hemlock, western	spruce	ponderosa	eastern
maple, sugar	lauan	red	spruce	pine
pine	almon	Sitka	redwood	eastern white
Caribbean	bagtikan	sweetgum	spruce	sugar
ocote	mayapis	tamarack	black	
pine, southern	red lauan	yellow poplar	Engelmann	
loblolly	tangile		white	
longleaf	white lauan			
shortleaf				
slash				
tanoak				

(a) Each of these names represents a trade group of woods consisting of a number of closely related species.

(b) Species from the genus *Dipterocarpus* are marketed collectively: apitong if originating in the Philippines; keruing if originating in Malaysia or Indonesia.

(c) Douglas fir from trees grown in the states of Washington, Oregon, California, Idaho, Montana, Wyoming, and the Canadian provinces of Alberta and British Columbia shall be classed as Douglas fir No. 1. Douglas fir from trees grown in the states of Nevada, Utah, Colorado, Arizona, and New Mexico shall be classed as Douglas fir No. 2.

(d) Red meranti shall be limited to species having a specific gravity of 0.41 or more based on green volume and oven-dry weight.

posed. Lower veneer grades are permitted for Interior plywood than for Exterior. Some Interior types are laminated with waterproof glue, but, because of the lower-grade plies, they are not as durable as Exterior.

Woods from more than seventy species of varying strength are used in plywood manufacture. They are divided into five groups. The strongest and stiffest are in Group 1; those weakest in these qualities are in Group 5. These standards are established by the American Plywood Association (APA) and apply to the species of face and back plies, or to the weaker of the two if they are different. Exceptions are decorative and sanded panels ⅜ inch thick or less; these are identified by the face species group.

Each type of plywood is available in a variety of grades, according to the veneer grade on the face and the back of the panel. Veneer grades define the appearance in terms of unre-paired growth characteristics and allowable numbers and sizes of repairs that may be made during manufacture. The top-of-the-line veneer grade is N, which is made only on special order. Grade A, with only minimal blemishes, is next. A-C and A-D plywood are sometimes referred to as *good-one-side*, while A-A is *good-two-sides*. (The first letter refers to the face ply, the second to the back ply.) The minimum grade of veneer permitted in Exterior plywood is C. Veneer graded D is used only for backs and inner plies of Interior plywood.

The type, group, and grade are stamped on the back or edge of all APA grade-trademarked plywood panels. Other information in the stamp identifies the mill where the panel was manufactured and the U.S. Product Standard to which it conforms.

For best results with painted finishes, Medium Density Overlaid

Plywood is made to precise sizes so use can be planned without waste.

Plywood panels are carefully graded by appearance of surface plies.

Veneer Grades for Softwood Plywood

N Smooth surface "natural finish" veneer. Select, all heartwood or all sapwood. Free of open defects. Allows not more than six repairs, wood only, per 4×8 panel, made parallel to grain and well matched for grain and color.

A Smooth, paintable. Not more than eighteen neatly made repairs, boat, sled, or router type, and parallel to grain, permitted. May be used for natural finish in less demanding applications.

B Solid surface. Shims, circular repair plugs, and tight knots to 1 inch across grain permitted. Some minor splits permitted.

C Improved C veneer with splits limited to 1/8 inch width and knotholes and borer holes limited to 1/4 × 1/2 inch.

Plugged Admits some broken grain. Synthetic repairs permitted.

C Tight knots to 1½ inch. Knotholes to 1 inch across grain and some to 1½ inch if total width of knots and knotholes is within specified limits. Synthetic or wood repairs. Discoloration and sanding defects that do not impair strength permitted. Limited splits allowed. Stitching permitted.

D Knots and knotholes to 2½ inch width across grain and ½ inch larger within specified limits. Limited splits allowed. Stitching permitted. Limited to Interior grades of plywood.

(MDO) plywood is recommended. MDO has a special resin-treated wood fiber surface permanently bonded to the panel under heat and pressure; this provides an excellent base for paint. MDO panels are available factory-primed.

Softwood plywood can be found at virtually every building supplies dealer in the country, in thicknesses of 1/4, 3/8, 1/2, 5/8, and 3/4 inch. Many lumberyards sell panels smaller than the standard 4 × 8 feet, and most dealers will also cut the panels to a customer's specifications, usually for a small charge.

Hardwood plywood

Hardwood plywood is the frequent choice where appearance is of primary importance, as for wall paneling and fine furniture. It is available with face veneers of a wide variety of wood species, both domestic and foreign. Some of the more exotic of these are very expensive, but they bring a richness and beauty to a project that is difficult to match.

In manufacture, hardwood plywood differs somewhat from the softwood variety. Most common is the veneer-core panel, with inner plies of a less expensive grade than the face veneer, and often thicker than in softwood plywood. *Veneer-core plywood* can be bent to a curve, its radius depending on the thickness of the panel. *Particleboard-core plywood* has the face veneers laminated to particleboard and is used in such applications as cabinetry. It offers good dimensional strength and stability. *Lumber-core plywood* has a solid wood core of strips of lumber edge-glued together. A crossband layer of low-grade veneer is bonded to both sides of this core, with the grain running at right angles to that of the lumber. The face and back veneers are then added, with grain at right angles to those layers beneath. Lumber-core plywood is widely used for table tops, cabinets, and doors.

Hardwood plywood is classified according to the glues used in its construction. *Type I* is bonded with waterproof glue and is designated for exterior use. *Type II* has a water-resistant bond and is for interior use; most furniture is made from Type II. *Type III* is moisture-resistant; it will withstand normal indoor dampness and humidity, but it can be damaged by water.

Premium Grade allows only a few small burls, occasional pin knots, slight color streaks without sharp contrasts, and inconspicuous small patches. For most species with prominent grains, multi-piece faces must be book-matched or slip-matched. *Good Grade* also proscribes sharp contrasts in color and great dissimilarity in grain and figure of adjacent pieces of veneer, but it does not consider matching at veneer joints. *Sound Grade* is smooth-surfaced, but veneers need not be matched for grain or color; they must be free of open defects. *Utility Grade* allows knotholes with diameters up to 1 inch, and wormholes and splits not exceeding 3/16 inch wide and extending less than half the length of the panel. *Backing Grade* permits knotholes up to 3 inches across and splits up to 1 inch wide, depending on the length of the split.

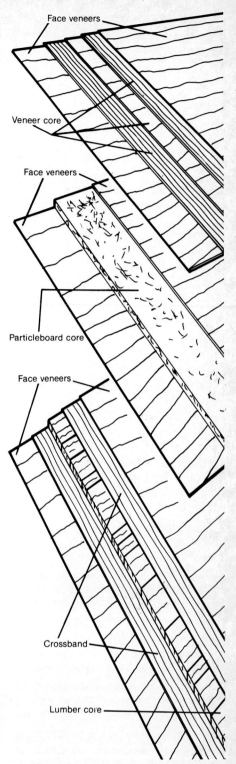

Hardwood plywood types: (top) veneer core, (center) particleboard core, and (bottom) lumber core.

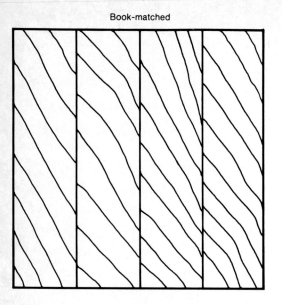
Book-matched

Slip-matched

Matching face-veneer patterns on plywood should be selected for the look desired in a project.

Hardwood plywood is available in thicknesses from 3/16 inch to 1½ inches; panel size is 4 × 8 feet, although the lumberyard may sell it in smaller sizes. Many manufacturers offer factory-finished panels. Building supply dealers often carry a selection of hardwood plywood or can order panels for you; but if you are looking for a particular species, especially one of the more rare types, you may have to shop around.

Hardboard

Hardboard (often called Masonite, although that is the trademark of one manufacturer of the product) is more dense, has more strength, and is more durable than a comparable panel of wood. This is due to its carefully engineered manufacture. Logs are reduced to chips, then the chips to wood fibers. The fibers are rearranged and permanently bonded together with lignin (the natural wood-bonding material). As in the tree itself, cellulose provides strength. Hardboard is impact-resistant and does not crack, splinter, check, craze, or flake.

There are three basic classifications of hardboard: Service, Standard, and Tempered. *Service* is intended for miscellaneous interior uses, such as closet liners, shelf coverings, cabinet backs, and drawer bottoms. *Standard* is for interior paneling (usually prefinished), underlayment, screen dividers, wardrobe

drawers, door skins, and cabinets. *Tempered* hardboard undergoes a special process in which oil is introduced into the panel and permanently set by heat. This gives the panel greater strength, abrasion resistance, and moisture resistance. Tempered hardboard is used for such exterior applications as soffits, shutters, flower boxes, garden sheds, and windbreaks.

Hardboard is commonly available in thicknesses of 1/8 and 1/4 inch, and in panels up to 4 × 8 feet. Larger panels may be ordered. The panels are smooth-surfaced on one or both sides. The smooth face provides an excellent base for paint.

Perforated hardboard (often referred to as Peg-Board, another tradename, although other manufacturers make similar products) is very useful for storage in garages, children's rooms, workshops, kitchens, utility rooms, and the like. The many precision-spaced holes in the panel can be fitted with a wide variety of hooks and hangers.

Particleboard

Another engineered wood panel product is *particleboard*, which consists of wood residues from lumber and plywood manufacturing operations—mainly planer shavings and veneer clippings. These are combined with synthetic resins and a wax emulsion (for moisture resistance) and hot-pressed into panels of

uniform thickness and size. The result is a strong, smooth, dimensionally stable panel. Like hardboard, it is without knots or grain, but it is less dense and comes in thicker panels.

Standard particleboard thicknesses are 3/8, 1/2, 5/8, and 3/4 inch; panels up to 1½ inches thick are available for heavy-duty industrial applications. Standard panel size is 4 × 8 feet.

Particleboard has many uses, often as a substitute for plywood. It serves as core stock for some hardwood plywood and as underlayment for resilient flooring. It may be used for drawers, bookcases, countertops, cabinets, and similar applications. It is an excellent shelving material; many dealers carry precut particleboard shelves in various lengths and widths.

Particleboard can be worked with ordinary tools, but the bonding agent used in its manufacture is very abrasive and rough on saw teeth. A carbide-tipped blade should be used for extensive cutting operations. It also does not hold nails well, and it tends to chip if nails are driven near panel edges. Screw-fastening is a better method, or adhesives may be used.

Particleboard panels take paint well, but they may need filling first, especially along the edges. A sealer is recommended under clear or stained finishes.

Appearance Characteristics of Hardwoods

Species	Frequency of knots	Color of heartwood*	Type of figure: In plainsawed lumber	In quartersawed lumber
ash:				
black	low	moderately dark grayish brown	conspicuous growth ring; occasional burl	distinct growth ring stripe; occasional burl
white	low	grayish brown, sometimes with reddish tinge	same as black ash	same as black ash
aspen	low	light brown	faint growth ring	none
basswood	low	creamy white to creamy brown, sometimes reddish	same as aspen	same as aspen
beech	intermediate	white with reddish tinge to reddish brown	same as aspen	numerous small flakes to 1/8" high
birch	low	light brown to dark reddish brown	distinct but not conspicuous growth ring; occasionally wavy	occasionally wavy
cherry	low	light to dark reddish brown	faint growth ring; occasional burl	occasional burl
cottonwood	low	grayish white to light grayish brown	faint growth ring	none
elm:				
rock	intermediate	light grayish brown, usually with reddish tinge	distinct but not conspicuous growth ring, with fine wavy pattern within each ring	faint growth ring stripe
slippery	intermediate	dark brown with shades of red	conspicuous growth ring, with fine pattern within each ring	distinct but not conspicuous growth ring stripe
hackberry	low	light yellowish or greenish gray	conspicuous growth ring	same as slippery elm
hickory	intermediate	reddish brown	distinct but not conspicuous growth ring	faint growth ring stripe
locust	intermediate	golden brown, sometimes with tinges of green	conspicuous growth ring	distinct but not conspicuous growth ring stripe
magnolia	low	light to dark yellowish brown with greenish or purplish tinge	faint growth ring	none
maple	low to intermediate	light reddish brown	faint growth ring; occasionally bird's-eye, curly and wavy	occasionally curly and wavy
oak:				
red	low	grayish brown, usually with fleshy tinge	conspicuous growth ring	pronounced flake; distinct but not conspicuous growth ring stripe
white	low	grayish brown, rarely with fleshy tinge	same as red oak	same as red oak
sycamore	low	flesh brown	faint growth ring	numerous pronounced flakes to 1/4" high
tupelo	low	pale to moderately dark brownish gray	same as sycamore	distinct but not pronounced ribbon
walnut	low	chocolate brown, occasionally with darker, sometimes purplish streaks	distinct but not conspicuous growth ring; occasionally wavy, curly, burl	distinct but not conspicuous growth ring stripe; occasionally wavy, curly, burl, crotch, other figures

* The sapwood of all species is light in color or virtually white unless discolored by fungus or chemical stains.

Appearance Characteristics of Softwoods

Species	Frequency of knots	Color of heartwood*	Type of figure: In slash-grained lumber	In vertical-grained lumber
cedar:				
eastern red	high	brick red to deep reddish brown	occasionally, streaks of white sapwood alternate with heartwood	occasionally, streaks of white sapwood alternate with heartwood
northern white	intermediate	light to dark brown	faint growth ring	faint growth ring stripe
southern white	low	light brown with reddish tinge	distinct but not conspicuous growth ring	none
western red	low	reddish brown	same as southern white cedar	faint growth ring stripe
cypress	low	light yellowish brown to reddish brown	conspicuous, irregular growth ring	distinct but not conspicuous growth ring stripe
Douglas fir	intermediate	orange-red to red; sometimes yellow	conspicuous growth ring	same as cypress
fir:				
balsam	high	nearly white	distinct but not conspicuous growth ring	faint growth ring stripe
white	intermediate	nearly white to pale reddish brown	conspicuous growth ring	distinct but not conspicuous growth ring stripe
hemlock:				
eastern	intermediate	light reddish brown	distinct but not conspicuous growth ring	faint growth ring stripe
western	intermediate	same as eastern hemlock	same as eastern hemlock	same as eastern hemlock
larch	high	russet to reddish brown	conspicuous growth ring	distinct but not conspicuous growth ring stripe
pine:				
Idaho white	high	cream to light reddish brown	faint growth ring	none
lodgepole	high	light reddish brown	distinct but not conspicuous growth ring; faint pocked appearance	none
northern white	high	cream to light reddish brown	faint growth ring	none
ponderosa	intermediate	orange to reddish brown	distinct but not conspicuous growth ring	faint growth ring stripe
southern yellow	low	same as ponderosa pine	conspicuous growth ring	distinct but not conspicuous growth ring stripe
sugar	high	light creamy brown	faint growth ring	none
redwood	low	cherry to deep reddish brown	distinct but not conspicuous growth ring; occasionally wavy and burl	faint growth ring stripe; occasionally wavy and burl
spruce:				
eastern	high	nearly white	faint growth ring	none
Engelmann	high	same as eastern spruce	same as eastern spruce	same as eastern spruce
Sitka	intermediate	light reddish brown	distinct but not conspicuous growth ring	faint growth ring stripe

* The sapwood of all species is light in color or virtually white unless discolored by fungus or chemical stains.

2 HAND TOOLS

It's a worn but still apt cliché: woodworkers are only as good as their tools. Perhaps more important, the converse is also true: tools are only as good as the person using them. You should take pride in your tool collection and in your ability to use those tools properly to do the jobs for which they are intended. There are tools for measuring, laying out, testing and checking, cutting, forming, smoothing, holding, boring, and fastening wood. Some are hand tools, which we will discuss in this chapter; others are electric, which we cover in the next two chapters.

The most valuable tools of all, however, won't be found in your toolbox. They are your hands. Whenever you use any other tools, protect your hands and the rest of your body—including your eyes—by handling the tool in the prescribed manner and taking all precautions that may be applicable to the woodworking operation being performed. Always follow *all* safety rules for the tool being used. This is just as important with hand tools as with power tools. The following safety cautions and shopping tips apply to the portable and stationary power tools discussed in the next two chapters as well as to hand tools. Read them carefully.

The right tool for the right job should always be your watchword.

Safety first—and always

One cardinal safety rule is to use tools only for the intended purpose. A screwdriver is *not* a chisel, nor is a wrench a hammer. For that matter, not all hammers are intended for all striking purposes. It is inefficient as well as dangerous to attempt to drive a spike with a tack hammer or a tack with a sledge. Select the right tool for the job at hand and learn—through practice—to use it correctly.

Keep your mind on what you are doing, no matter how skilled you may become in tool usage. Many an experienced carpenter has lost a finger or more because his attention wandered while he was using a power saw. Think safety at all times. This applies to hand tools as well as power tools—anyone who has inadvertently smashed a thumb with a hammer (and is there a woodworker who hasn't?) can attest to that.

Here are a few basic safety rules to follow.

Keep cutting tools sharp. This helps them to do their jobs better, of course, but it is also a safety precaution. A dull tool is much more likely to slip off the work and mar the surface or even injure the user.

Never use damaged tools. A battered screwdriver may slip and spoil the screw slot, damage the work surface, or cause painful injury. A split hammer handle may break when a nail is struck, causing the hammer head to become a dangerous flying missile. Don't just wrap tape around such a split; replace the handle or the hammer.

Keep the work area clutter-free. If

A battered hammer like this one is a potential hazard—discard it.

you are using portable power tools, make sure that the cords are long enough for the job, that they do not pose tripping hazards, and that they are out of the way of the tool's action.

Watch out for other people, especially children, in the work area. If youngsters seriously want to learn, teach them. Lesson number one is respect for tools, and that means giving a wide berth to the person using them. Let the children watch, at a safe distance. And don't be distracted by conversation, with children *or* adults, while you are working. That's an invitation to injury.

Dress for the job. Especially when you are using power tools, avoid loose-fitting clothing that can be caught up by a power saw or drill. For operations where there is a possibility of flying chips or debris, put on safety goggles. Corrective lenses, if you wear them, are no substitute. Purchase safety goggles that can be worn over your glasses. Your eyes, like your hands, are tools far too valuable to risk.

Shopping for tools

Often, the neophyte woodworker, fascinated with his newly discovered craft and the tools that make it all possible, will outfit his workshop with every gadget and gimcrack to be found at the hardware store and in the mail-order catalogs. That's an expensive and unnecessary approach to carpentry and woodworking. You will need a few basic tools to start (more on that below), but beyond that, it's best to buy tools only as the need for them arises. It makes no sense, economically or otherwise, to clutter up your work area with exotic devices capable of performing woodworking wizardry far more complex than you would ever want to attempt.

What tools you buy for starters will depend on what type of woodworking you intend to do. A bare-bones toolbox for around-the-house carpentry will contain a claw hammer, a crosscut saw, one or two screwdrivers, a small jack plane, a pair of pliers, an 8- or 10-foot flexible tape measure, and a ¼-inch portable

power drill with a set of drill bits ranging from ¹/₁₆ inch up. You can build on that base as you need specific tools for special purposes and as your budget allows.

You don't have to spend a fortune on tools, but don't buy cheap either. Bargain-basement specials are seldom satisfactory and often hazardous to use. Never try to get by by selecting lower-priced tools that aren't really the type or size you need. If you are going to do the job, you might as well do it right—and that means using the right tools.

Examine each tool carefully before you buy it. Look for a sturdy body and a smooth finish. Metal surfaces should be coated for rust prevention; wood parts should be varnished, waxed, or lacquered for durability and protection against splinters. Check all moving parts to make sure they work smoothly and easily but are free of any play or wiggle. Look for tools permanently marked with the manufacturer's name or symbol as an indication of careful construction and quality material. Ask about warranties on power tools; some quality hand tools are also warrantied.

Tool storage

"A place for everything and everything in its place." A tool is useless if you can't find it. If you have a workshop, a wall of perforated hardboard makes a fine storage place—just make sure you hang up the tools after you use them. A good toolbox is also a good storage place and allows you to take the tools where they are needed. But don't just throw tools into the toolbox. First clean them, then set them gently into the box. Careless handling can dull cutting edges, cause nicks to striking faces and other metal components, and damage moving parts. Once again, it's a matter of respect and common sense.

Types of hand tools

Woodworking was probably born when a Stone Age savage discovered that a sharp piece of flint could be used to carve a point on the end of

a stick. Very likely, several millennia passed before another of our ancestors realized that it was easier to pound that pointed stick into the ground with a rock than with his fist, and thus the hammer was invented. Through the ages, countless other tools were developed as the need arose—all of them devices to extend man's use of his hands.

Examples of beautiful work created by man and his hand tools are abundant: the highly ornamented armchairs found in the tomb of King Tut in Egypt; the fine veneer work and inlays on furniture of the Roman Empire; the ornate carvings in the great cathedrals of the Middle Ages; the elegant cabinets and chests of the Renaissance; the stylistic masterpieces of Thomas Chippendale, George Hepplewhite, and Duncan Phyfe; the classic mansions built by colonial carpenters; the sturdy simplicity of Shaker designs; the lacy gingerbread of Victorian architecture. All were crafted with essentially the same hand tools in use today. It is, once again, a matter of selecting the right tool for the job and learning to use it correctly.

Hammers

Every handyman's toolbox should contain at least one hammer, a *carpenter's hammer*. This is made in two patterns: curved claw and straight (or ripping) claw. The claws are for pulling or withdrawing nails. Curved claws are better suited to this purpose; however, straight claws can

Curved claw hammer with wood handle.

Straight claw or ripping hammer.

also serve to pry apart nailed-together pieces, which is more difficult to do with curved claws.

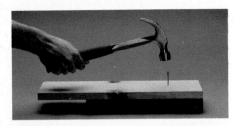

Hold the hammer near the end of the handle to take advantage of the weight of the head.

When pulling long nails, place a block under the hammer head to provide extra leverage.

Another consideration is the weight of the hammer head. Claw hammers range from about 8 ounces up to 28 ounces (the larger sizes are also called framing hammers). For general use, a 16-ounce carpenter's hammer is recommended. If you will be doing only small workshop projects, a 12-ounce hammer may suffice.

The carpenter's hammer is a simple tool, and just about anybody who can lift one should be able to master its use. But it does take some practice. The most common error is "choking" the hammer—holding the handle too close to the head. This reduces the leverage and the force of the blow and makes it more difficult to hold the head flat against the surface being struck. If a nail is not hit squarely by the hammer's striking face, it is likely to bend. Grasp the handle firmly, close to the end, with your fingers underneath and your thumb alongside or on top of the handle (your thumb should be on the handle, not overlapping your fingers). For very light blows, you can swing the hammer with your wrist. Otherwise, swing with your elbow, or your wrist and your elbow. For heavy

work, such as driving spikes, get your shoulder into the action too. Just remember that the striking face of the hammer and the surface being struck should be as near parallel as possible.

If you are driving a nail into a finish surface, lighten the blows as the hammer nears the surface, and stop before the nail head reaches the surface. Use a nailset slightly smaller in diameter than the head of the finishing nail to complete driving the nail to the surface or below, so that the hole can later be filled with wood putty.

To withdraw a nail whose head is above the surface, slip the hammer claws under the head as far as they will go, then pull back on the handle. If it is a long nail, place a block of wood under the head of the hammer for extra leverage. If the nail head is flush with or below the surface, place the ends of the claws around it and tap the face of the hammer with a soft-faced mallet (never with another hammer) to drive the claws around the head. Then pull back on the handle. If the nail is in a finish surface that you do not wish to mar by driving in the hammer claws, use a nailset to drive the nail completely through the piece of wood, rather than trying to withdraw it. You can later patch the hole with wood putty.

As your tool collection grows, it will probably include hammers other than the basic carpenter's claw hammer. A *tack hammer* is a good investment for light-duty jobs: driving tacks and brads. Usually weighing about 5 ounces, the tack hammer may have a wood, steel, or fiberglass handle. When you shop for one, make sure that it has a magnetic head (either the striking face or the opposite end). This is a finger-saving feature, especially handy for driving short tacks and brads. The magnetic head holds the tack while you tap or press it into the wood, then you can drive it in.

Soft-faced mallets of wood or plastic have cylindrical heads with two flat striking faces; handles are usually wood or fiberglass. They are used primarily for striking wood- and plastic-handled chisels and gouges

and to form sheet metal. They can also be used to drive a claw hammer over a nail to be pulled (as described above).

A *mash hammer* is a heavy, short-handled tool with a double-faced striking head. A *ball-peen hammer* has a rounded striking face with beveled edges and a ball-shaped peen, intended for machine- and metal-working applications rather than woodworking, at the other end of the head. Always wear safety goggles to protect your eyes when using these hammers. Chips all too frequently fly off from steel chisels, especially if they are struck off center, and nail heads may break off the hardened nails.

Position the nailset directly over the head of the nail and drive the head below the surface of the wood.

A magnetic tack hammer—for small jobs.

A soft-face mallet for wood chisels.

A mash hammer for all-steel chisels.

woodworking, at the other end of the head. Always wear safety goggles to protect your eyes when using these hammers. Chips all too frequently fly off from steel chisels, especially if they are struck off center, and nail heads may break off the hardened nails.

Saws

The familiar *carpenter's handsaw* consists of a steel blade with a handle, most commonly wood, at one end. The blade is narrower at the end opposite the handle. This end is called the point or toe; the end of the blade at the handle is the butt or heel. The lower edge of the blade has cutting teeth, which are set (bent alternately side to side) to make the kerf (the groove cut by the teeth) wider than the thickness of the blade. This is necessary so that the blade does not rub against the sides of the kerf and bind. In addition, good-quality saws are taper-ground, with the top edge thinner than the toothed edge; this further reduces friction inside the kerf.

Carpenter's handsaws are described by the number of teeth per inch. (Actually, since the measurement is made from point to point on the teeth, there is always one more point in the designation than there are teeth per inch.) A number stamped on the blade near the handle gives the point size of the saw. In general, the lower the point size, the easier and faster the saw will cut; the higher the point size, the finer and smoother the cut.

There are two broad classifications of carpenter's handsaws: ripsaws and crosscut saws. *Ripsaws* are designed to cut with the grain of the wood; *crosscut saws* cut across the grain. The latter should probably be your first purchase.

The major difference between a crosscut saw and a ripsaw is the shape of the teeth. On a crosscut saw, the teeth have beveled, knifelike cutting edges that sever the wood fibers running with the grain of a piece of wood. For general use, an 8-point crosscut saw is recommended. If the saw will be used primarily for fine work—cutting trim or hard-

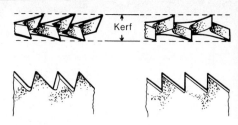

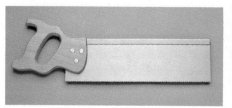

Kerf is the gap cut by saw teeth: crosscut (left), rip (right).

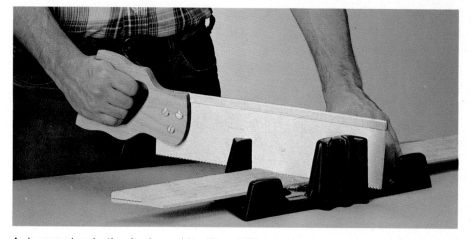

The backsaw has a heavy steel spine.

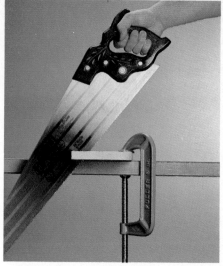

A crosscut saw should be held at about 45° to the surface of the work.

An inexpensive plastic miter box guides 45° and 90° cuts. Here it is used with a backsaw.

woods—a 10- or 12-point may be preferable.

Ripsaws have teeth with square-faced, chisellike cutting edges that do a good job of cutting with the wood grain, but a poor job of cutting across it. Most ripsaws have 5½ or 6 points per inch.

When using a carpenter's handsaw to cut a piece of wood, place the board on a sawhorse or similar sturdy object, or in a vise. For long boards, use two sawhorses so that the board rests solidly. Use a square and a pencil or an awl to mark a cutting guideline on the wood. Hold the saw in your right hand (unless you are a southpaw) and grasp the board with your free hand. For crosscutting, hold the saw at an angle of approximately 45° to the wood surface; for ripping, hold it more nearly vertical, about 60°. Be sure that the

side of the saw blade is *plumb*—at right angles—with the face of the board.

Although these saws are designed to cut on the push stroke, it is usually easier to start the cut by placing the heel of the blade on the cutting line and pulling it toward you to begin the kerf. Then take short, light strokes—applying pressure on the push stroke only—and gradually increase the strokes to the full length of the saw. Keep the action smooth; do not force or jerk the saw. Your sawing arm should swing clear of your body so that the saw handle is at your side rather than in front of you. Keep your eye on the guideline, not on the saw, while cutting; this will enable you to see any tendency of the saw to wander. If it does, take shorter strokes and twist the handle slightly to bring the saw back to the

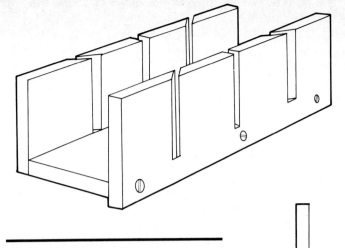

straight and narrow. Blow away saw-dust frequently so that the guideline is not obscured. Slow down to make the final strokes of the cut. Hold the waste piece in your free hand so that the board will not split as the last few strokes are made.

If a saw sticks or binds during a cut, it may indicate a dull or poorly set blade. Other causes are too much moisture in the wood or forcing the saw. You can use a wedge in the kerf behind the blade to prevent binding; this will probably be necessary when ripping long boards.

A *back saw* is a crosscut saw designed for cutting a perfectly straight line across a piece of wood. A heavy steel backing, or "spine," along the top of the blade keeps it perfectly aligned. Back saws usually have 12 or 13 points per inch. The saw may be used freehand; but for more precise cuts, use it in conjunction with a miter box. The miter box ensures exact 90° or 45° cuts. Many commercial miter boxes can be adjusted for other angles, or you can make your own miter box out of wood, slotted for whatever angle cuts a project requires (see *Make your own Miter Box*).

For very fine cutting, a *dovetail saw* is the choice. This looks something like a back saw, except that the handle is in line with the steel backing along the top of the blade (which is narrower than that of the back saw). The dovetail saw usually has about 15 points per inch. As the name implies, this saw is designed to make the fine cuts for close-fitting wood joints, but it can also be used with a shallow miter box. Its cutting operation is similar to that of the back saw.

A *compass saw* is used for making curved or internal cuts—for example, a switch opening in a piece of paneling. Its blade is narrower at the toe than at the heel and has 8 or 9 points per inch. A *keyhole saw* is much the same as a compass saw; it is slightly smaller and has 10 or 11 points per inch, making it better suited to finer work.

Techniques for using both these saws are the same. When you are cutting in from the edge of a piece,

Make your own miter box

Use nominal 1-inch hardwood stock for the miter box. The length and width of the box are optional, and you may size the box to fit the work you intend to do. Minimum inside width should be 4 inches; minimum length 16 inches. The height of the sides above the bottom cannot be more than the width of the back-saw blade measured from the bottom of the teeth to the steel spine. Note that one side of the box extends about 1 inch below the bottom, so that the box may be gripped in a vise or braced against the edge of a workbench or saw-horse.

Assemble the box with glue and wood screws. Use a combination square to mark the 90° line and two 45° lines (one in each direction) across the top edges, then carry these lines down the sides. (90° and 45° angles are the most common cuts made in a miter box, but if your project calls for any other angle, simply mark that angle on the box.) With the back-saw, carefully cut along the lines down the sides to the base. Accuracy in these initial cuts is essential, since all other cuts made will follow the same angles.

start the saw with a few pull strokes, then take longer push strokes for cutting. To cut curves, twist the handle to guide the blade. Since the blade is narrower at the toe, you can make tight-radius curves by shortening the strokes and using only the toe of the blade. For internal cuts, drill a starting hole first. Then insert the toe of the blade and take short strokes until the kerf is long enough for full strokes to be made. Very long inside cuts may be started with a compass saw; when the kerf is long enough,

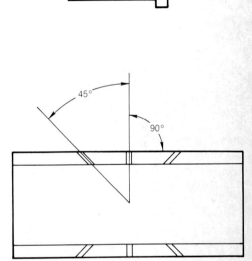

The secret to making a good miter box is measuring the cutting angles accurately.

insert a crosscut saw to finish the job more easily.

A *coping saw* consists of a thin, fine-toothed (10 to 20 points per inch) blade held in a U-shaped metal frame. The blade is removable and can be inserted in the frame so that it cuts on either the push or pull stroke in any direction. Some more expensive types of blades are round, with spiral cutting edges; these will saw in any direction, making possible tight-radius curves. To make an internal cut with a coping saw, remove the blade from the frame and pass it through a pre-drilled hole in the work; then reattach the blade to the frame for the cut. When using a coping

- Always keep saw blades sharp. A dull blade not only makes you work harder, but may also stick or jump out of the kerf, marring the work and possibly causing injuries.
- Never allow the teeth of a wood saw to come in contact with concrete, stone, or metal. Before sawing a piece of wood, examine it carefully to make sure that there are no nails in the path of the cut.
- Set the saw down gently after use—never just throw it aside or into a toolbox where it can come in contact with other tools.
- If possible, store the saw by hanging it up where there is no chance of its teeth being damaged. If you must carry a saw in your toolbox, hold it in a special rack or compartment where the teeth are protected.
- Sharpening a saw is a skill in its own right—a time-consuming one as well. Most craftspeople prefer to leave this job to the professional who has the proper tools to do it right. Many hardware stores offer saw-sharpening services. If yours doesn't, look in the Yellow Pages of your phone book under "Saws—Sharpening & Repairing."
- It's a good precaution to protect the saw blade by giving it a coat of paste wax. If the blade begins to rust, clean away the rust by rubbing it with a fine emery cloth. Then coat the blade with a light machine oil.

saw, hold the work firmly in a vise, as close to the cutting line as possible, to avoid the possibility of splitting and excessive blade twisting and possible breakage.

A hacksaw is not a woodworking tool, but you may need one if you have to cut through metal as part of a carpentry or woodworking project. A *hacksaw* has a replaceable blade

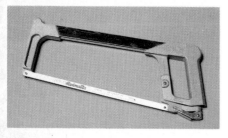

Adjustable-frame hacksaw.

held in a frame that is either solid or adjustable. Adjustable frames can be used with blades of various lengths, while solid frames accommodate only one length. Hacksaw blades have from 14 to 32 points per inch; the lower point sizes are used for heavy materials, while 32-point blades do the best job on thin sheet metal. The blade is usually installed in the frame with the teeth pointing away from the handle so that it cuts on the push stroke. A wing nut or other device in the frame is tightened to hold the blade taut, helping to make straight cuts.

If possible, secure the material to be cut in a vise—but not your woodworking vise. If you must use your woodworking vise, protect the jaws by sandwiching the metal between blocks of scrap wood. It is difficult to guide the saw when the piece cannot be placed in a vise, and you have to hold it with your free hand.

Assuming that the work is secured in a vise, grasp the saw handle in one hand and, with your other hand, hold the outer edge of the hacksaw frame. Start the kerf as you would with any other saw: take a few short back strokes. Then take longer strokes, applying pressure on the push stroke and easing off on the return stroke. Take long, steady strokes, letting your body sway ahead and back with each stroke. Don't apply so much pressure that the blade bends. When you reach the end of the cut, slow down to prevent the blade from snagging in the kerf. Support the waste end so that it doesn't break off and leave a jagged edge.

Drills

Most wood-boring operations today are performed with a power drill, but the traditional hand tools are still preferred by many craftspeople, who feel that hand tools do neater and finer work than the power variety. You will probably have both hand and power types in your tool collection.

A *push drill* can be used to drill holes up to ¼ inch in diameter. It consists of a handle, a shank with an enclosed spindle, and a chuck to

Push drill with fluted drill point.

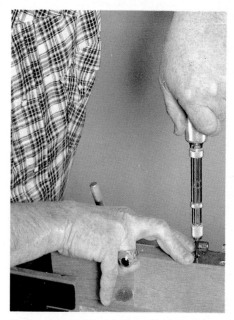

Push, then the handle springs back.

Hand drill or "eggbeater" drill.

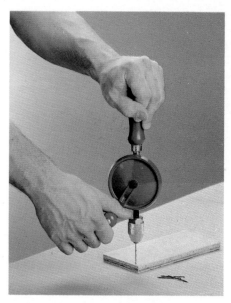

Using a hand drill for a pilot hole.

hold a drill point. When the handle is pushed, the spindle rotates the chuck. When pressure is relaxed, an internal spring returns the handle to the original position for the next push. Many push drills have hollow handles with screw caps to hold a supply of drill points in sizes from $1/16$ inch to $1/4$ inch.

The push drill uses a straight, fluted drill point. To chuck the drill, loosen the chuck several turns and insert the drill as far as it will go. Turn the drill until it seats in the driving socket in the bottom of the chuck, then hand-tighten the chuck to hold the drill point securely.

If possible, secure the work in a vise or with clamps. If you must hold the work in place while drilling, do not grasp it too close to the drilling location. Take care to make the hole as perfectly horizontal or vertical as possible. It often helps to make a starter hole with an awl. Then place the drill point in the starter hole and operate the push drill with alternate push-pull strokes.

A *hand drill* (eggbeater) is also used to drill holes up to $1/4$ inch in diameter. It is operated by turning a geared crank, which then rotates a shaft with a chuck holding the drill— either a twist drill or a drill point as is used in a push drill. Like the push drill, a hand drill often has a magazine inside the handle to store a variety of points. Drills are chucked the same as in a push drill.

Since the operation of this drill requires both hands, the work must be secured in a vise or by clamping. Use one hand to turn the crank while guiding the drill and applying slight pressure on the handle (which is opposite the chuck end of the tool) with your other hand.

With a *brace and bit*, you can bore larger and deeper holes than with either a push drill or hand drill. The brace is a crank with a chuck on one end to hold a drill bit. Better quality braces have an adjustable ratchet mechanism that allows you to use back-and-forth motions for boring operations, a feature that is especially desirable in cramped working conditions.

Brace with adjustable ratchet.

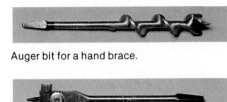

Auger bit for a hand brace.

Auger bit with interchangeable cutters.

The auger bit is the tool that is chucked in the brace to actually do the cutting. Auger bits have tapered-end shanks to fit into the brace. At the other end of the shank is a twist. At the end of the twist are two sharp points called spurs, which score the circle in the wood, and two cutting edges, which cut shavings within the scored circle. A screw on the tip of the bit centers the bit and draws it into the wood. As it penetrates, the twist carries the cuttings away from the cutters and deposits them around the hole.

The most common auger bit set ranges in diameter from $1/4$ inch to 1 inch, in $1/16$-inch increments. The bit sizes are stamped on the tang (the part that enters the chuck): a number 12 means $12/16$, or $3/4$, inch; number 5 means $5/16$ inch, and so on. Ordinary auger bits are from 7 to 9 inches long. Shorter bits, 3 to $3 1/2$ inches long, are called dowel bits.

Expansive auger bits have adjustable cutters for boring holes of different diameters. These bits usually come with two interchangeable cutters and will bore holes with diameters from $1/2$ inch to 3 inches. A scale on the cutter blade indicates the diameter of the hole to be bored.

To chuck an auger bit in a brace, hold the shell of the chuck and turn the brace handle to open the jaws. When the jaws are far enough apart,

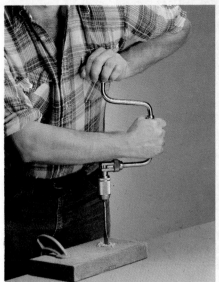

The brace is steadied against the body as the crank is turned.

insert the square tang of the bit until the end seats in the square driving socket at the bottom of the chuck. Then tighten the chuck by turning the handle to close the jaws and hold the bit in place.

Use a sharp pencil to mark the exact center location of the hole to be bored. Then make an indentation in the wood at that point by pressing in an awl. Place the point of the feed screw on the bit in the indentation and, if possible, steady the brace against your body, with the auger bit square with the surface of the work. Apply only slight pressure as you turn the crank to bore the hole.

To prevent splintering as the bit breaks through the wood, clamp a block of scrap wood behind the work piece. Another way to bore through without splitting the back face is to reverse the bit when the feed screw just becomes visible through the opposite face. Pull it out of the hole and finish the hole by boring from the opposite face—the hole made by the feed screw serves as center point.

Wood chisels

A *chisel* is a wood-cutting tool consisting of a steel blade fitted with a hardwood or plastic handle. The blade has a single beveled cutting edge on the end. There are two general classes of chisels: *tang chisels*, in which part of the blade is enclosed

Start a chisel cut between saw kerf cuts by marking the deepest point of the cut with a light tap on the chisel.

Chip out the waste wood between the kerf cuts, starting near the edge and working back in small pieces.

Finish the cut by cleaning out rough edges and splinters with light taps on the chisel and finally by hand.

by the handle, and *socket chisels*, in which a tapered end of the handle fits into a matching socket on the blade. A socket chisel is designed for striking with a soft-faced mallet. A tang chisel is meant primarily to be worked by hand, although it will withstand some light blows.

Tips on chisel use

- Secure work in a vise or by some other means so that it cannot move.
- Keep both hands behind the chisel's cutting edge at all times.
- Do not start a cut on a guideline. Start slightly away from it so that there is a small amount of material to be removed by the finishing cuts. This ensures a clean cut.
- When starting a cut, always chisel away from the guideline toward the waste wood so that no splitting will occur at the edge.
- Never cut toward yourself with a chisel.
- Don't cut too deeply. Keep the shavings thin, especially when finishing.
- Whenever possible, cut with the wood grain; this severs the fibers and leaves the wood smooth. Cutting against the grain splits the wood and leaves it rough. This type of cut is also difficult to control.
- Protect the chisel's cutting edge during use and storage. A dull or nicked chisel cannot produce a smooth cut. It forces you to apply more pressure, increasing the possibility that it may slip off the work, damage the surface, or cause serious injury. A dull tool is a dangerous tool.

Wood chisels are also categorized according to weights and thicknesses, and to the shape or design of the blades. *Paring chisels* have light, thin blades about 2½ inches long. *Pocket chisels* have heavier blades about 4½ inches long. *Firmer chisels* are all-around tools with heavier weight blades about 6 inches long. *Butt chisels* have standard-weight blades, but the blades are about 3 inches long and are used in work where longer chisels would be awkward. Common chisel widths are, in inches, ¼, ⅜, ½, ⅝, ¾, 1, 1¼, 1½, and 2. They are often purchased in sets; a good beginner's set might include ¼-, ½-, ¾-, and 1-inch widths.

Whenever possible, other tools—saws or planes—should be used to remove most of the waste from the work piece; use the chisel for finishing purposes only. Secure the work firmly in a vise. For rough cuts, hold the beveled edge of the chisel against the work and force either by hand pressure or by tapping with a mallet. For smoothing and finishing cuts, hold the chisel with the flat side or back of the blade against the work; hand pressure supplies the driving power. The blade should not be pushed straight through the opening, but should be moved laterally at the same time that it is pushed forward. This ensures a shearing cut, producing a smooth and even surface even when the work is across the grain.

Carving chisels are specialized tools designed primarily for use by hobbyists in handcrafting articles from soft woods. They are made in a wide variety of configurations and are usually purchased in sets. Among the most common types are the *straight chisel*, for straight whittling; the *bent chisel*, for cutting concave surfaces; the *skew chisel*, for carving fine detail; the *gouge*, for cutting out a round, concave surface; the *"U" chisel*, for grooving and cutting concave surfaces; and the *veining chisel*, V shaped for cutting fine lines or veins. These tools must be kept very sharp, so extra caution should be taken when using them.

Planes

When you buy lumber for your projects, it is normally dressed, that is, smooth on all four sides. If no further cutting is required (except for length) final smoothing before finishing is usually done by sanding. But when a small amount of the wood must be removed for a fit, a plane will do the job. This tool is also used to smooth the edge after a board has been ripped. Planes are also used to shave doors, sashes, and drawers that bind.

Bench planes and *block planes* are designed for general surface smoothing and squaring and are the planes most familiar to do-it-yourselfers. A bench plane has a handle at the rear that is grasped to push the plane ahead and a knob at the front that is held to guide the plane along its course. The main body of the plane is called the frame and has an opening in the bottom through which the blade, or plane iron, protrudes. A plane iron cap is screwed to the upper face of the blade to deflect shavings upward, preventing the opening from becoming choked with jammed shavings.

Bench planes come in various lengths and weights, but all are used primarily for shaving and smoothing with the grain of the wood. A smooth plane is about 9 inches long; a jack plane, 14 to 15 inches long; a jointer plane, from 20 to 24 inches long. The longer the plane, the more uniformly flat and true the planed surface will be. The smooth plane is, for all but the smallest work, a smoother only; it will plane a smooth but not especially true surface. It can also be used for cross-grain smoothing and squaring of board ends. The jack plane is the best all-purpose tool if you will have only one bench plane in

The jack plane is a good, all-around tool for general shop use.

The plane iron cap fits over the blade, or plane iron, to deflect shavings upward through bottom.

your collection. It can take a deeper cut and make a truer surface than the smooth plane. The jointer plane is used when longer surfaces must be as true as can be achieved, as when two boards are to be edge-joined.

The plane iron is set at the correct angle and desired depth by means of an adjusting nut and an adjustment lever. Generally, the blade should be set to cut only a thin shaving from the wood. If much material is to be removed, the blade can be set to make a thicker cut, then retracted so that the final cuts are very thin to ensure smoothness.

The work to be planed should be held in a vise (or otherwise firmly braced). Always plane in the direction of the wood grain; otherwise the wood will split and splinter. If it is necessary to plane against the grain, set the blade to make as thin a cut as possible. Hold the plane securely, with one hand on the handle and one on the knob; start the pass by placing the front of the tool (the toe) squarely on the end of the work. Apply light and even pressure as you push the tool along the work, and do not let up until the plane has cleared the other end.

A block plane is a small (usually 6 or 7 inches long), single-handed plane that is cupped in the palm and pushed by the heel of the hand. The blade in a block plane does not have a plane iron cap, as does a bench plane, and the blade goes in bevel-up —the reverse of a block plane iron. It is also set at a lower angle.

The block plane, which is usually held at an angle to the work, is especially suited for squaring end grain and smoothing plywood. It is also useful for very small work.

A *spoke shave* is a specialty plane designed to smooth or round off concave or convex shapes. Most provide for blade adjustment so that the depth of cut can be controlled. The spoke shave has handles at each side. To use it, grip the tool with both hands and exert pressure downward, in the direction of the pass on the work. As with other planes, the thinner the cut, the smoother the surface.

The block plane has a blade at a low angle and no plane iron cap.

A block plane is cupped in the palm and usually held at an angle.

The spoke shave is useful for smoothing curved shapes.

Plane blades must be kept very sharp if they are to do satisfactory work. Before storing a plane, always retract the blade so that it cannot be damaged. If you are planing old wood or wood that has been previously used, check for nails before you begin. If possible, avoid planing knots; this can damage the cutting edge of the plane iron.

Formers

Not all woodworking tools boast a long lineage. Relatively new on the scene are *formers* (Surform® tools by tradename), versatile tools that can shape, shave, file, plane, and smooth. They come in several shapes and sizes, with replaceable blades, for one- or two-hand operations. Each blade has hundreds of hardened, tempered, and razor-sharp preset steel teeth. Each tooth is a tiny chisel, ground to a fine edge. The blades resemble a cheese grater, and the cutting action is similar. Stand-

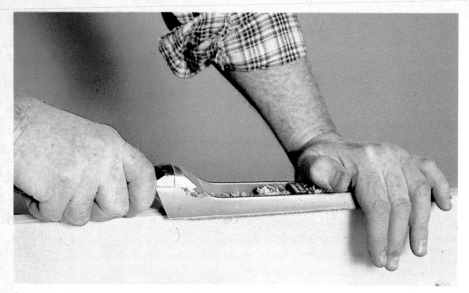

A plane-type former requires two-handed operation, much like a bench plane. Each of the hundreds of teeth is like a tiny chisel.

The shaving-type former cuts on the pull stroke.

The round former shapes decorative cuts and enlarges holes.

The former should be held at an angle to the surface to remove wood quickly, then gradually brought parallel to the surface for smoothing.

ard blades handle most jobs; fine-cut blades are best for end grain, dense woods, or very fine work. Round blades work on curved surfaces; narrow blades are useful for slotting; wide blades are best for broad, flat surfaces.

To remove a maximum amount of material with a former, hold the tool at a 45° angle to the direction of the stroke and apply pressure as you move it across the surface. Reduce the angle and the pressure as you near the end of the cut. To remove only a small amount of material and produce a smooth surface, simply direct the tool parallel to the work surface.

Screwdrivers

A *screwdriver* is designed for one purpose only: to drive and remove screws. Unfortunately, many do-it-yourselfers—and professionals too —use screwdrivers for chiseling, prying, scraping, scoring, and even punching holes. The tool lends itself fairly well to all these applications, but such abuse usually renders it unfit for its task of driving screws. Again, always use the correct tool for the job at hand, and for that job only!

Screwdriver safety

- Do not hold work in your hand when using a screwdriver—if the point slips, it can cause a bad cut.
- Whenever possible, hold the work in a vise, with a clamp, or on a solid surface. If this is impossible, follow this rule: never have any part of your body, including your hands, in front of the screwdriver tip.
- Make sure that the blade tip fits the screw slot.
- Never try to turn a screwdriver with a pair of pliers. For heavy-duty work, use a wrench on a square-shank screwdriver.
- Never pound on a screwdriver.
- Do not use a screwdriver as a pry bar.
- Keep the blade, shank, and handle in good condition. Blade tips that become nicked or burred can usually be redressed with a grinding wheel or metal file. If a handle or shank is damaged, discard the tool.

Never use a screwdriver that doesn't perfectly fit the screw slot—and of course never substitute another tool for a screwdriver. The screwdriver shown at the top is too small and might damage the screw slot. The one at the bottom is correct.

Conventional screwdrivers consist of a handle, usually of wood or plastic, with a steel shank extending from it, culminating in a flat blade that fits into a slotted screw. The steel shank is designed to withstand considerable twisting force in proportion to its size, and the tip of the blade is hardened to keep it from wearing.

Standard screwdrivers are classified by size, according to the combined length of shank and blade. The common sizes range from 2¹⁄₂ inches to 12 inches, and the well-stocked workshop will have a complete set within that range. If you want to build up toward that ideal, start with 3-, 6-, and 8-inch screwdrivers and add as the need arises.

There are many variations of the standard screwdriver. Stubby versions are intended for use where working space is constricted. Cabinet screwdrivers have long, thin shanks with narrow blades; they are useful for driving screws into recessed and counterbored openings in furniture. Some screwdrivers for heavy-duty work have square shanks so that they may be gripped with a wrench for turning. This is the only type of screwdriver on which a wrench should be used. Never use pliers to turn a screwdriver; this will only chew up the shank.

Recessed-head screws have a cavity formed in the head and require specially shaped screwdrivers. There are several types of such screws, but by far the most common used in woodworking is the Phillips head screw. This has an X-shaped slot into which the screwdriver fits, preventing the tool from slipping and damaging the slots or the wood surrounding the screw. Three standard-size *Phillips head screwdrivers* handle a wide range of screw sizes; always select the size that fits the recess snugly, otherwise the tip will be damaged and its effectiveness destroyed.

A pilot hole should always be made before driving a screw, especially in hard woods or when a screw is near the end of a board. For small screws, press an awl into the wood to make the pilot hole. For larger screws, and in dense hardwoods, drill a pilot hole. Insert the tip of the screw in the pilot hole, then insert

The conventional screwdriver.

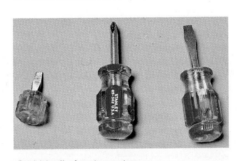

"Stubbies" of various sizes.

Cabinet screwdriver.

Phillips head screwdriver.

Spiral-ratchet screwdriver.

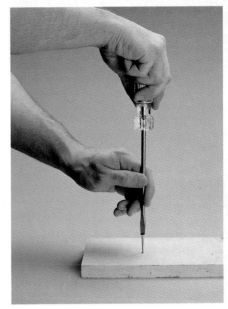

When driving a screw, keep the shank of the screwdriver in line with the shank of the screw to avoid driving the screw out of its intended line or damaging the screw slot with the blade.

the screwdriver tip into the screw slot or recess. Grip the handle firmly with one hand, making sure that the shank of the screwdriver is perfectly in line with the shank of the screw, and hold the tip steady with the other hand. Turn the handle to drive the screw, steadying the shank until the screw is almost all the way in. You can then use both hands, if necessary, to turn the handle, applying pressure to the top until the screw is fully seated.

For fast and easy work, especially when driving screws in soft woods, a *spiral-ratchet screwdriver* is very convenient. It does not require that the blade be lifted out of the slot after each turn; you simply press it down its full length, then release pressure and a spring returns the shank to its original position.

Wrenches

You don't need an extensive array of wrenches for most woodworking projects; these are more the tools of a mechanic or machinist. Still, you should at least have an adjustable *open-end wrench* in your toolbox. One jaw of this tool is fixed, while the other is moved along a slide by a thumbscrew to fit any size nut within its opening capacity. Adjustable

The adjustable open-end wrench is a versatile tool for any shop.

wrenches are available in sizes ranging from 4 to 24 inches in length. An 8-inch wrench, with jaws opening to $^{15}/_{16}$ inch, is a good basic choice.

When using an adjustable wrench, always make sure that the jaws are snug on the nut. Position the wrench so that the nut is all the way into the throat of the jaws, then pull the handle toward the side with the adjustable jaw. Failure to do this may result in the wrench slipping off the nut, rounding its square or hexagonal corners, and possibly bloodying your knuckles.

An adjustable wrench can also be used to turn a square-shank screwdriver when driving a screw into very dense wood. Adjust the jaws to fit snugly on the shank, then apply pressure to the screwdriver handle while turning the shank with the wrench.

Measuring tools

Accurate measurements are essential for any successful woodworking or carpentry project, so it's important to select and use measuring tools with care. For most purposes, a flexible steel tape is the popular choice.

Steel tapes come in many lengths, and in widths of $^1/_4$, $^3/_8$, $^1/_2$, $^3/_4$, and 1 inch. For general use, a 10- or 12-foot tape is usually recommended. Shorter tapes (6-foot, 8-foot) are inadequate for such jobs as measuring full sheets of plywood and hardboard; longer tapes are bulkier and, unless absolutely needed for frequent use, a waste of money. The tape is usually curved across its width, which gives it a certain amount of rigidity when extended without other support. At the end of the tape is a hook that fits over the edge of the board or panel being measured, so that the job can be done without a helper to hold the end. For taking inside measurements, the hook slides into a slot in

the tape to ensure complete accuracy.

The tape is contained in a metal or plastic case. Better quality tapes retract automatically after use; others need to have the tape pushed in after use, which poses the hazard of kinking the steel. Many also include a lock to hold the tape in any extended position, lessening the possibility of error.

Common graduations are in inches, with lines representing sixteenths and eighths of inches. Flexible tapes are also available in metric and metric/inch graduations.

When making a measurement with a tape, make sure that it is perfectly straight and at a right angle to the edge or end from which the measurement is being taken. Look straight down at the point where the measurement is to be marked; if you

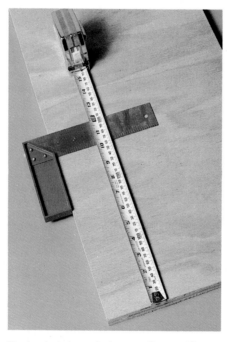

The hook at the end of a tape fits over the end of the work.

For inside measurements, the hook slides into a slot on the tape.

look down at an angle, you are likely to misread the measurement.

To take an inside measurement, such as the inside of a box, place the case inside the box, snug against one side, then pull out the tape until it is tightly butted against the other side, with the hook retracted. Read the tape at the point where it enters the case, then add the width of the case (usually 2 or 3 inches).

Measuring tips and tricks

- For every measuring operation—and in this we include marking angles and other measuring-related steps—repeat the sequence of measurements to ensure that you get the same result both times. If they do not agree, recheck, and re-mark if necessary.

- Whenever possible, try to avoid measuring—use a precut length or width of material as a guide, or hold the piece of stock to be cut in place against the parts that are completed and then mark and cut it to fit. If you are experienced at cutting all components to size beforehand, by all means do so; you will save time. For others, however, the cut-to-fit method offers less chance of wasted material.

- Another trick is to add up the component sizes for a particular item and then compare it to the overall size. The two should be the same, after deducting overlaps or joints from the sum of the components. For example, if you were installing a series of cabinets and knew that they should occupy 6 feet of wall space, the linked-together widths of the cabinets to be installed should also equal 6 feet.

- If using a ruler, try to measure without using the first inch of the tool. The metal clip on the tool often makes the rule difficult to read. If you have had the tool for some time, the edges might be damaged and thus slightly inaccurate. It is safest to read the length beginning at the 1-inch mark or even at the 2-inch mark.

- If working with someone else on a project, be sure that you both use the same brand of ruler. There are usually fractional differences between ruler brands —12 feet ends up a different actual dimension for different manufacturers— and this can result in mismatched pieces of precut, expensive materials.

Many zigzags include extensions for inside measurements.

Using a combination square to check that the end of a piece of stock is square.

For very long measurements, steel tapes are available in lengths of 50 and 100 feet. These tapes must be manually cranked.

Folding wood rules (also called *zigzag rules*) are favored by many carpenters. Metal joints lock the rule in open position; when extended, it is rigid enough to span openings. The most common length is 6 feet, although longer and shorter models are available. Like the steel tape, it comes in inch, metric, and metric/inch graduations. Many folding rules include a sliding brass extension for taking inside measurements.

Squares

Squares are essential for laying out work, for marking angles and checking for trueness after cutting boards and paneling, and for checking corners during assembly, whether you are framing a house or building shelves. Take care never to drop these tools, and do not store them in toolboxes where they may get banged around by other tools. If the angle between the blade and the handle is distorted, the square is useless.

A *try square* is used for checking or marking surfaces that must be at right angles to each other. It consists of two parts at a perfect right angle to each other: a steel blade and a handle of wood, plastic, or metal. The blade may be marked in 1/8-inch or 2-millimeter graduations. Blades run from 6 to 12 inches long; handles, from 4 to 8 inches long. For all-around use, an 8- or 10-inch blade is best. Some try squares have handles cut at 45° angles where they join the blade. Sometimes called miter squares, these can be used to check both 90° and 45° angles.

More versatile than the try square for general home workshop use is the *combination square*. It consists of a grooved steel blade, 12 inches long, with a handle that slides along the groove. A thumbscrew in the handle locks it at any position on the blade. The handle is shaped so that it can be used to mark or check either 90° or 45° angles. It can also be used to check the depth of holes or slots. The blade is placed in the hole, then the handle slid down to the surface of the work and locked. The blade reading gives the depth of the hole.

The handle of the combination square is removable, so that the blade can also be used as a flat ruler. Many models have leveling vials built into the handle; these tools can be used for determining plumb (perfectly vertical) and level (perfectly horizontal). Some also include a small, removable scratch awl in the handle.

The *carpenter's* or *framing square*

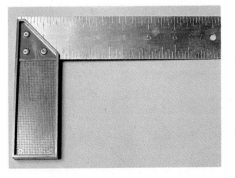

The try square is used to mark and check right angles.

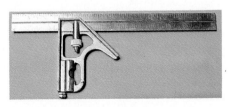

The combination square has a movable handle that locks into place.

is a one-piece, L-shaped tool of steel or aluminum. In the most common type, the longer leg of the L (the body or blade) is 24 inches long and 2 inches wide, while the shorter leg (the tongue) is 16 inches long and 1 1/2 inches wide. The 24- and 16-inch lengths greatly simplify the layout of framing members with 24- and 16-inch o.c. spacing; the 1 1/2-inch width of the tongue is the same as the width of 2x dimension lumber, another aid in layout work. Both body and tongue are graduated in inches and fractions of an inch.

The most frequent uses for this square are laying out building framing, squaring up large patterns, and testing the flatness and squareness of large surfaces. Squaring is accomplished by placing the square at right angles to adjacent surfaces and observing if light shows between the work piece and the square.

The carpenter's square also has a wealth of information stamped on it to simplify or eliminate the need for computations in many woodworking tasks. Depending on the specific purpose for which the tool is intended, these markings may include metric conversions; fraction-to-decimal conversions; board-foot equivalents; formulas for laying out six- or eight-sided figures; 30°, 45°, and 60° angle markings; and screw and nail sizes. A variation, the *rafter square*, is stamped with rafter tables, taking the mystery out of the angle cuts required on rafters for roofs of various pitches.

Levels

Levels are precision tools designed to check and prove whether a plane or surface is true horizontal (level) or true vertical (plumb). While you probably won't need a level for most home workshop projects, such jobs as structural carpentry or building shelves and built-ins require frequent checks with a level.

A level consists of a glass or acrylic vial partially filled with a liquid (usually alcohol) mounted in a frame of wood, aluminum, or magnesium. On the outside of the vial are two lines separated by a space cor-

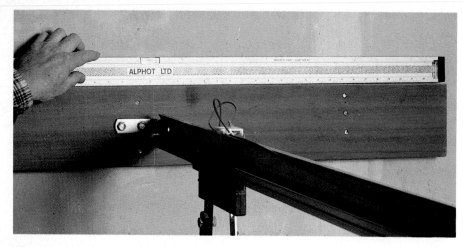

When checking that something is level, eye the level's vial directly to see if the air bubble is centered between the lines.

Hold the level tightly against the surface and note the air bubble.

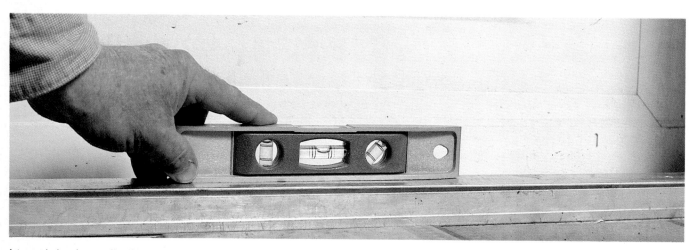

A torpedo level—smaller than a carpenter's level—is the best tool for checking smaller work. This one has vials for level and plumb, plus a third vial set at 45° angle for checking diagonals.

responding to the width of the air bubble trapped inside the vial. Level is achieved when the air bubble is centered between the lines. Most levels contain at least two such vials mounted at right angles to each other, so that the tool can be used for both plumbing and leveling.

For general purposes, a 24-inch *carpenter's level* is a good choice. To level a piece of work, a shelf, for example, set the level on the shelf parallel to the front edge. Eye the level from directly in front of the vial, and note the position of the air bubble. If it is not perfectly centered between the lines, raise one end of the shelf and recheck the bubble. When you are satisfied that it is perfectly level, fasten the shelf.

To plumb a piece of work, such as a framed wall, place the level verti-

cally on the face of the work and check the bubble in the appropriate vial. Adjust the work as necessary until the bubble is centered.

For smaller work, a *torpedo level* comes in handy. A shorter version (usually 8 or 9 inches long) of the carpenter's level, it contains vials for both plumb and level; some include a 45° vial as well for checking that angle in relation to a vertical or horizontal surface.

As a precision instrument, a level should be respected. Take care not to drop it or handle it roughly. Protect the edges from damage and store it in a rack or other suitable place when not in use.

Spot-check the level occasionally for accuracy. To do this, place it on a horizontal surface and make a pencil mark at each end of the frame. Note

the position of the bubble. Now reverse the level end-for-end and align it with the pencil marks. If the bubble is in the same position, the level is true. If the bubble is in a different position, the level is out of true and the vial must be adjusted. (How this is done will depend on the make and model of your level, so keep the instructions that come with your level.)

Chalk

A chalk line is frequently used to mark a straight line between two points, such as a guideline for building a partition on a floor deck during house construction. It may simply be a cord rubbed with colored chalk, then drawn tautly between the two points to be joined and snapped in the center to transfer the chalk mark

Snapping a chalkline is a quick way to make long marks.

A pry bar, shorter than the ripping bar, pulls hard-to-reach nails.

A ripping bar, also called a crow bar or wrecking bar.

ity, such as a bump, could throw off the string measurement.

Stand in front of the string after it stops swinging, and pencil in two marks: one directly behind the string at the ceiling; the other behind the string near the bottom of the wall. Move back a few feet and, keeping

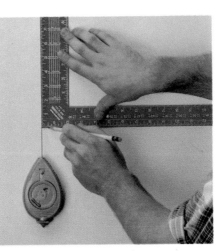

One way you can use a square is to find a true horizontal measuring from the vertical line provided by a plumb bob.

to the surface beneath. For more frequent use, a *chalk line reel* is much handier. The line is automatically chalked as it is pulled from the reel case; after use, it is retracted by a crank. Some reel cases come to a point at one end; these can double as plumb bobs to locate a point directly beneath a point overhead or to mark an exact vertical line. The cord is secured to the upper point with the case just over the lower point. When the case stops swinging, the bottom of the case will be directly below the upper point.

Plumb bobs

A *plumb bob* is a teardrop-shaped weight attached to a string. To use it, just hang the plumb bob from the ceiling about an inch away from the wall. Make sure the string does not touch the wall; any surface irregular-

one eye closed, sight up and down the string. The string should obscure both marks. If it does not, try again. Then snap a chalkline across the two marks for a straight line.

Bars

For construction projects, a variety of ripping bars and pry bars can be very useful for gaining extra leverage, prying, pushing, lifting, and pulling stubborn nails. A ripping chisel, although technically a wood chisel,

seems more at home in this category, since it can also perform many of those functions; it can also be struck with a ball-peen or mash hammer for heavy-duty chiseling.

Vises

A vise is used for holding work while it is being sawed, bored, planed, formed, or glued. The familiar bench vise is designed for metalworking and so is not suited for working with wood. A *woodworker's* or *carpenter's vise* is the tool you should have. The jaws on these vises are large enough to protect the wood and to provide a good grip on larger pieces. Many woodworker's vises have a rapid-action feature that allows the movable jaw to be moved in and out quickly, with the final tightening accomplished by giving the handle a half-turn or so.

Larger vises should be attached to the workbench with bolts or heavy

This chalkline reel case is pointed so it can double as a plumb bob.

This vise has a convertible jaw and can be screwed into place.

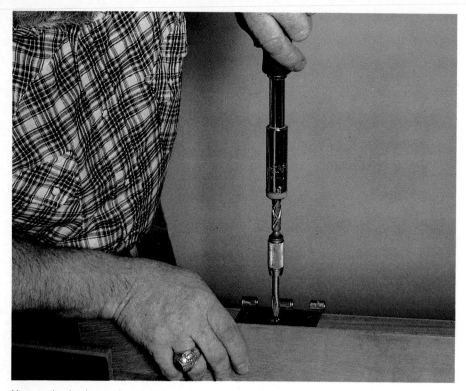

Here, a vise is clamped to a sawhorse to hold a door while a hinge is attached with a spiral ratchet screwdriver.

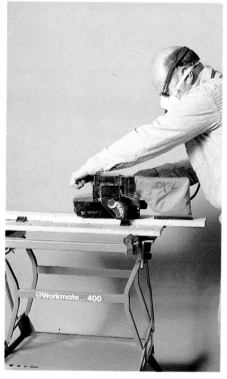

This combination vise, workbench, and sawhorse folds for storage.

lag screws. The vise should be mounted so that the top of the jaw is flush with the surface of the workbench and flush with the corner of the bench. Some smaller vises have screw clamps allowing them to be mounted on benches, sawhorses, or other supports.

Vise advice

- Never use a vise that is not firmly secured or clamped to a workbench or other support.
- Never use an extension handle ("cheater") on the vise handle for extra tightening pressure.
- Never pound on the handle to tighten beyond hand pressure.
- When clamping extra-long work, support the far end of the work rather than putting extra pressure on the vise.
- When work is held in the vise for sawing, saw as close to the jaws as possible to prevent vibrations. Do not cut into the jaws.
- Do not clamp work between damaged jaw liners. Touch liners up with sandpaper or replace them if they are in bad shape.
- Do not let sawdust and dirt accumulate on the threaded part of the vise.

Metal vise jaws are lined with hardboard or hardwood to protect the work. Slight scars or gouges that may develop in the lining material can usually be removed by sanding. If the liners become badly worn or seriously damaged, replace them with similar material.

A specialty tool particularly useful in a workshop of limited size is the *work center*. A combination vise/workbench/sawhorse, it can be carried to a job site, and folded and stored when not in use. Several accessories—clamps, cutting guides, shaping guides—are available for use with the work center.

Clamps

Clamps are used when work cannot be satisfactorily held in a vise because of the size, shape, or type of project. There are as many sizes, shapes, and types of clamps as there are special needs for them; but, as with any tools, buy them only as your projects require. For general purposes, a few basic clamps should serve.

Most common is the *C clamp* (so called because of its shape,

although many say it should be called a *G clamp* for that reason), consisting of a steel frame threaded at one end to receive an operating

Clamping do's and don'ts

- It should go without saying, but it needs to be said: keep your fingers out of reach of the clamps you are tightening.
- Store clamps in a rack; do not place them in a drawer or loose in a toolbox.
- Use pads, scrap blocks of wood, or both between clamps and the work to avoid marring the surface.
- Never use a wrench, pipe, hammer, or pliers to gain extra leverage when tightening a wood clamp.
- Avoid using large clamps simply because of their deep throats; always fit the clamp to the job.
- C clamps are meant for clamping. Never use them for hoisting or for supporting scaffolding.
- Discard any clamp that has a bent frame or a bent threaded spindle. It cannot do the job for which it was intended.
- Keep all moving parts lightly oiled and clean. Make sure there is no oil or dirt on any part that will come in contact with the work.

screw with a swivel head. A wing nut or a sliding-pin handle is used for tightening and securing the work. The swivel head makes the clamp self-aligning without digging into the work, but finish surfaces should be protected by wood scraps to avoid marring. C clamp capacities (the size of the work that can be clamped) range from 1 to 12 inches. Larger clamps are not suitable for light-duty work, nor should you expect a small clamp to exert heavy pressure. Consider the required opening and the strength and the weight needed, then select a clamp to match the job.

C clamps can also be used effectively to glue up corner joints. With paper sandwiched to permit easy removal, glue triangular blocks near the end of each piece to be joined and let the glue set. Apply glue to the mitered ends and pull together with C clamps, making sure that the pieces are in perfect alignment. After the glue has set, remove the C clamps, pry away the blocks, and sand off the paper.

A variation of the C clamp is the *bar clamp*, which consists of a steel bar with a fixed jaw at one end and a sliding or adjustable part containing a screw with a handle on one end and a swivel head on the other. A series of notches in the bottom of the bar makes it possible to slide the movable part close to the work and be engaged; the handle is then turned to tighten the clamp.

Bar clamp capacities range from 6 inches to 6 feet, depending on the length of the bar. This clamp is very useful for gluing up extra-wide work.

The *pipe clamp* is a close relative of the bar clamp. It is used with 1/2-inch or 3/4-inch plumbing pipe, threaded on one end. The pipe is screwed into a fixed jaw, and an adjustable jaw is installed on the other end of the pipe. The movable jaw is slid along the pipe to the work, where it locks in place. A handle or crank is then turned for final tightening. The capacity of the pipe clamp is limited only by the length of pipe available.

A *web clamp* is a heavy-duty nylon or canvas strap wrapped around the

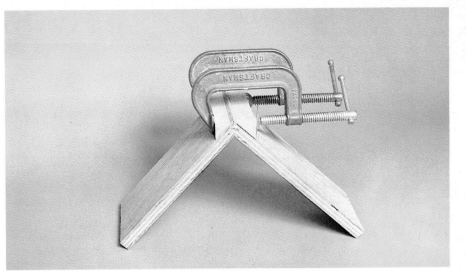

C clamps are indispensable in any home workshop. Here, with angled wood blocks, they hold a mitered corner fast while glue sets.

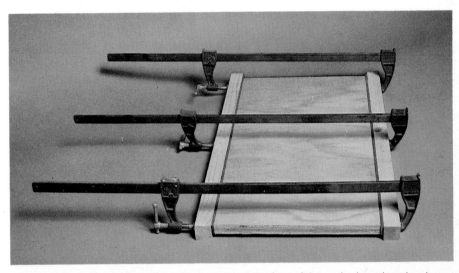

Bar clamps handle the bigger jobs. Always protect the face of the work when clamping; here scraps of wood are used between clamp jaws and surface.

Handscrews should be tightened so the full length of the jaws contact the surface. The clamp at the right is askew and will not apply even pressure.

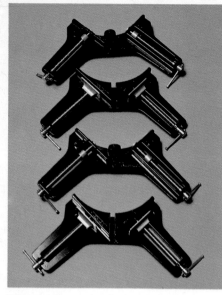

Spring clamps apply light-to-moderate holding pressure and are good clamps to use when fast application of pressure is desirable.

Miter clamps apply simultaneous pressure to all four corners of a frame while glue is drying.

work to be glued or held together during assembly. By means of a ratchet, the strap is drawn tightly to apply the required pressure. This clamp is useful for gluing up large work, such as the rungs of chair or table legs, and for gluing irregularly shaped assemblies.

The *handscrew* consists of two hardwood jaws connected by two operating screws, one with a left-hand thread and the other with a right-hand thread. The jaws move in opposite directions due to the action of the opposing threads as the handles are turned. Handscrews are used primarily to hold work together during gluing. The smooth wood jaws protect the finish of the work by spreading the clamping pressure over a broad surface. The jaws can also be adjusted to assume a non-parallel position, if necessary to match the surfaces of the work. Handscrew capacities range from 2 to 12 inches.

Miter or *corner clamps* are used to apply simultaneous pressure to all four joints of a square or rectangular frame while gluing. It is important to test for squareness at each corner before applying final pressure, and to apply equal pressure at all corners.

Spring clamps are handy when fast application and removal of pressure is desirable, but they can also be used where light to moderate holding power is required for gluing or other operations. Spring clamps range from 4 to 9 inches in length, with jaw capacity from about 7/8 inch to 4 inches. Some types are available with vinyl-padded jaws to protect the work.

3 PORTABLE POWER TOOLS

Electricity can supply the muscle for many woodworking operations. Portable power tools are relatively inexpensive, and precise controls allow you to produce highly accurate work. But, as with hand tools you must use the tool correctly and match the tool to the job.

The advantages of portable power tools are almost evenly distributed between their power and their portability. With an electric motor supplying high-speed operation to a power tool, a job can go much faster than when using hand tools. While the fine woodworking hobbyist may take pride in his handwork, most professional carpenters and cabinetmakers use power tools because "time is money."

The portability of a power tool is often essential: if you need to work on your roof, you can hardly bring the roof down to your shop. Of course, for many jobs, portable power tools are used as if they were stationary in the shop with only their price and capacity for certain kinds of work distinguishing them from their larger cousins.

Power tool safety

Read and understand the owner's manual *before* using any power tool, and keep the manual handy for reference. Use the tool only for those jobs it is intended to perform and never try to coax it beyond its capabilities. Follow the manufacturer's recommended maintenance procedures.

Unless the tool is double-insulated, it should be plugged into a three-hole grounded outlet. Don't use an adapter plug to connect a three-prong plug into a two-hole outlet; it may not provide adequate grounding. Without grounding, an electrical malfunction in the tool could cause serious—even fatal—shock. No matter how well the tool is insulated or grounded, never use it in wet or damp conditions. Moisture readily conducts electricity, posing severe hazards.

Heavy-duty tools (11 to 13 amps) should never be connected to an electrical circuit on which another appliance is operating. This could cause circuit overloading, blowing a fuse or tripping a circuit breaker. Do not replace a fuse with one of a higher amperage in order to carry the load—this is extremely dangerous and could cause overheated wiring and fire. Tools drawing 8 to 10 amps can be operated on household circuits if no major appliances are on the same circuit. Tools with lower amperage can be plugged into any household circuit not overloaded with several small appliances in operation.

Keep tool power cords away from heat, oil, and sharp edges that can damage the insulation. Don't carry a tool by the cord or jerk the cord to remove the plug from the electrical outlet; this may break the wire or its insulation. A damaged cord should be *replaced* (not repaired) immediately.

Guard against tool overheating. This may cause a breakdown of its electrical insulation and start a fire in the tool; at the least it will damage the tool and render it unsafe for further use. Keep housing vents clear so that air can flow freely inside the tool. If a tool does begin to heat up, turn it off immediately and allow it to cool. You may have been pushing it too hard, or the vents may have become clogged. Never wrap a cloth around a hot tool to hold it; this increases tool heat by blocking air flow, and the cloth itself may catch fire.

Do not operate any power tool near open containers of flammable solvents, cleaners, or varnishes. Fumes from such substances mix with air, forming a potentially dangerous explosive that can burst into flame from a motor or friction spark.

When changing bits, blades, cutters, or abrasives, first make sure that the tool switch is in the "off" position and disconnect the power cord from the electrical outlet. Then follow the directions in the owner's manual to make the change. When you are ready to use the tool, check to see that the switch is still "off" or that your finger is not on the trigger, then plug in the power cord.

Extension cords

An *extension cord* is often needed to get the power to the tool at the job site. The cord should be of a suitable gauge (wire size) for its length and the ampere rating of the tool. Otherwise, the power supply to the tool may be decreased by a "voltage drop" that can impair the tool's performance and may cause permanent damage to the motor.

A three-wire extension cord is required for use with tools having three-prong plugs. Two-wire cords are suitable for use with double-insulated tools.

When shopping for an extension cord, check the American Wire Gauge (AWG) stamp on the outer insulation. A 16-gauge, three-wire cord will be stamped "16/3," "16/3 AWG," or "3 conductor 16 AWG." The cord should also have a safety seal of approval from Underwriters' Laboratories (UL). This symbol indicates that the cord (or tool) has been manufactured to conform to established electrical and mechanical safety standards, although this does not represent a guarantee of the product. If the cord will be used outdoors, select one labeled "Suitable for use with outdoor appliances," or some similar statement. Any other cord will be unsafe.

Shopping for portable power tools

Visit a hardware or department store that carries a wide array of power tools. Price alone is not sufficient to guide your selection. Any wise shopper knows that cheapest is seldom best, but neither is a high price tag a guarantee of quality—or satisfaction.

Check the tool for the UL symbol. Don't buy a tool without this symbol. The power tool carton may also display a manufacturers' association seal, such as that of the Power Tool Institute. This shows that the tool not only meets safety standards but has been inspected under power at the factory and has instructions for safe use in the carton.

There are two kinds of motors used in electric tools: rotary and vibratory. A *rotary motor* is connected to gears and attachments to control power delivery. This type is quite durable, but it contains brushes that eventually wear, causing loss of power, excessive sparking, or motor

failure; the brushes must then be replaced. *Vibratory motors* are used in some jig (saber) saws and finishing (orbital) sanders, supplying direct back-and-forth motion. These motors contain no brushes and are best suited for light-duty work.

A power tool protects the user from electric shock by either of two systems. It may be externally grounded, with a wire that runs from the housing through the power cord to a third prong on the power plug, which is connected to a grounded, three-hole electrical outlet. The grounding wire will carry any current that leaks past the electrical insulation of the tool away from the user and into the ground of the house wiring. Another method of protection is double-insulation, in which an extra layer of electrical insulation on the tool eliminates the need for a grounding plug and outlet. Both systems are equally effective, so your choice may depend on the availability of three-hole outlets in your home or workshop. Put remember that these systems of insulation cannot be combined. You cannot use an adapter on a three-prong plug and expect the wire to be grounded.

Look on the manufacturer's nameplate for the amperage rating, that is, the quantity of electric current drawn by the tool during operation. Most models have ratings of from 2 to 13 amps. Generally, the larger the tool, the greater the amperage rating. If two models are otherwise equal, the one with the higher rating will provide more power.

The tool housing shields the motor, gears, and electrical parts from dirt; it also holds them in place. Housing materials commonly used are aluminum and plastic, both lightweight and sturdy. Some housings include portholes for inspection and easy replacement of worn or damaged motor brushes.

Check the tool's warranty before you buy. Some promise free repair of faulty parts and guarantee workmanship for one year. Others may guarantee replacement of a defective tool by the local dealer. Some have no

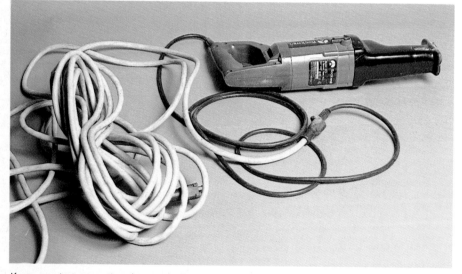

If you must use an extension cord with a power tool, make sure that it is adequate for the amperage rating of the tool.

Recommended Wire Gauges for Extension Cords

Ampere rating	Gauge			
	25-foot cord	50-foot cord	75-foot cord	100-foot cord
2 to 7	18	18	18	18
8 to 10	18	18	16	16
11 to 13	16	16	14	14

time limit as long as the original owner keeps the tool. But no company will provide free repairs if their examination indicates a tool failed because of abuse or unauthorized usage.

Electric drills

The versatile *electric drill* will probably be the first power tool you will buy. It can do much more than make holes. Its various attachments and accessories give it a wide range of capabilities. Electric drills also come in a wide range of prices, depending on quality, size, gearing, and special features.

Chuck size and rated revolutions per minute (rpm) determine the work capacity of a drill. The chuck size is the diameter of the largest bit shank the chuck can accommodate. Most common for do-it-yourself purposes are ¼-inch and ⅜-inch chucks; ½-inch chucks are also available. The rpm rating is an indication of the type

Drill with ½-inch chuck and handle.

of work for which a drill is best suited. Generally, the larger the chuck size, the lower the rpm rating. For woodworking, a ¼-inch drill rated at 2000 to 2200 rpm is appropriate; a ⅜-inch drill should run at about 1400 to 1500 rpm; a ½-inch drill, at 600 to 800 rpm.

For most woodworking applications, a single-speed drill is adequate. Two-speed or variable-speed models may be preferable if you will be using many accessories, or if you intend to drill materials that require a slow speed, as glass. A drill with both variable speeds and a reverse feature is useful for driving and removing screws. But keep in mind that all these extra features add to the price of the tool and may be an extravagance if you rarely or never use them.

The chuck is most commonly a three-jaw gear type. The chuck collar is first hand-tightened on the shank of a bit, then a key is inserted into the chuck to do the final tightening, securely holding the bit. Some bargain drills have chucks that are simply hand-tightened by turning a knurled collar. Avoid these. They generally have poor holding power, and it may be difficult to loosen them when work is finished. Never use a pliers or wrench to tighten or loosen a chuck.

Most electric drills have pistol-grip handles. Before buying, lift the drill to get its feel. It should be comfortable and well balanced and allow you to get a firm grip. Some models, especially in ⅜- and ½-inch sizes, have removable side handles so that the drill can be grasped by both hands when the work is heavy or in an awkward position.

The trigger switch that starts the drill is on the handle. Many models include a switch lock, allowing continuous operation without keeping a finger on the trigger. The lock is activated by pushing a button; squeezing the trigger instantly deactivates it. Variable-speed drills have trigger switches that allow you to vary bit speed from 0 to maximum rpm by finger pressure. Some have controls that preset maximum rpm. Drills with a reverse feature have separate reverse controls. To protect the motor and gears, always allow the drill to come to a full stop before reversing.

Electric drill accessories

Making holes is, of course, the primary purpose for which the electric drill is designed, and for this there is a variety of drill bits. Unlike the square-tanged shank of a brace bit, the shank of a power drill bit that is grasped in the chuck is round and smooth.

Most common is the *twist bit*, which has a sharp point and two spiral-shaped cutting edges that lift chips out of the hole as the bit turns. High-speed steel twist bits are suitable for drilling wood and soft metals. Diameters range from ¹⁄₁₆ to ½ inch, and, while they can be bought individually, they cost less if purchased in sets.

The *spade bit* cuts larger holes. It has a flat driving end with a pointed tip. Cutting diameters range from ⅜ to 1 inch. *Boring bits* have round cutting heads to drill holes in the same size range as spade bits.

For still larger holes, a *hole saw* may be chucked into the drill. This consists of a rim saw blade and a centered pilot bit. Common diameters range from 1 to 4 inches.

Drill with detachable side handle.

A wood screw *pilot bit* has three widths of cutting edge. The narrowest drills a pilot hole for screw threads. The next widest makes a shaft for the unthreaded screw shank. The widest makes a recess, or countersink, for flat head screws. A *counterbore bit* is similar, except that it drills the countersink below the surface of the work so that the screw head can be concealed by a wood plug.

A *screwdriving bit*, for either slotted or Phillips head screws, attaches to drills with variable-speed and reverse features to drive and remove screws. On single- or two-speed drills, a special screwdriving chuck must be used with these bits.

Wheels and discs for a variety of purposes are secured to the drill by an arbor adapter. One end of the arbor goes through a center hole in the wheel or disc and is fastened by a washer and nut or by a screw and washer. A flange keeps the wheel from slipping down the shank that fits into the chuck. Discs are used either with abrasive paper for sanding or with a soft bonnet for buffing and polishing. A wheel with abrasive strips (Sand-O-Flex® is the most widely available) makes short work of finishing flat or irregular surfaces.

A drill stand holds the drill in place while you feed the work to it, essentially making it a stationary power tool. The stand is permanently attached to a workbench, but it is designed so that the drill can be inserted or removed easily. A vertical stand holds the drill with the bit pointing down; a handle lowers the drill bit into the work, allowing the drilling of precisely spaced and positioned holes. A horizontal stand holds the drill for convenient positioning of work against revolving accessories, such as a sanding disc or buffer.

Circular saws

The *circular saw* is a major time- and labor-saver in building projects. It is also one of the most dangerous power tools, although it need not be if safety practices are scrupulously followed. More on that later.

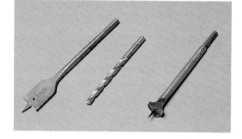

A spade bit is used for drilling holes up to 1 inch in diameter. A twist bit is most commonly used for drilling 1/16- to 1/2-inch holes. A boring bit is used for drilling holes from 3/8 to 1 inch in diameter.

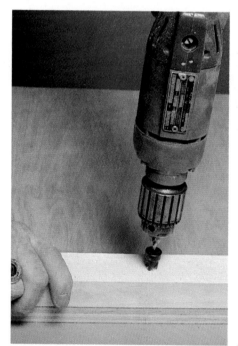

The wood screw pilot bit makes a recess for a flat head screw.

Polishing bonnet on a disc used with a right angle attachment on drill.

A drill-driven abrasive strip wheel can get into hard-to-reach spaces.

A bench top drill press can be easily stored.

Circular saw sizes are designated by the diameter of the blade. For do-it-yourself usage, these range from 6 to 8 or 9 inches. Amperage ratings are from 6 to 13. For frequent, heavy-duty cutting, a higher-amperage model is best. However, you can cut the same material with an economy saw if you work slowly and guard the motor against overheating.

The housing of a circular saw encloses the motor, electrical parts, and gears; it also incorporates blade guards, a base plate or shoe, cutting-depth and cutting-angle adjustments, and, usually, a removable ripping guide. There is a stationary upper blade guard covering the front, top, and back of the saw blade and a lower guard that covers the blade bottom (below the shoe) when the saw is not in use. The lower guard moves backward and upward as the saw is pushed into the work, then springs back automatically to cover the blade after the cut is completed. *Never* tamper with or remove this guard.

The cutting-depth adjustment moves the shoe up or down. As a rule of thumb, the blade should be set so that the gullets between the teeth just clear the bottom of the work. The cutting-angle adjustment tilts the shoe to any angle up to 45°. It includes a scale so that you can set the desired angle, and a wing nut or knob so that you can lock the shoe at that angle.

The work capacity of a circular saw is determined by its maximum depths of cut, both vertically and at a 45° angle. The depth of cut of a saw with a 6-inch blade should be at least 1¾ inches vertically and about 1½ inches at a 45° angle. Larger blades usually cut deeper. These depths of cut vary from manufacturer to manufacturer, so be sure to check the specifications before purchasing.

The ripping guide is an adjustable sliding scale at the side of the shoe that guides the saw blade parallel to the edge of the work. A wing nut or thumbscrew locks it in position.

A circular saw has a pistol grip for right-hand use; some models also include a front handle or knob for the left hand. The trigger switch is non-locking; power is turned off instantly when the trigger is released. A desirable feature found on some models is a slip clutch, designed to prevent motor burnout if the saw blade sticks in the work and to reduce the likelihood of kickback and possible serious injury.

There are several types of blades that can be used in a circular saw. For all-around use, a combination blade is the best choice. There are also crosscut blades, rip blades, and extra-fine blades to minimize tearing when sawing plywood and other panel materials. Always be certain that the tool is not plugged in when you change blades. Use the special wrench that comes with the saw to remove the screw and flange holding the blade. Attach the new blade, making sure the teeth point upward toward the front of the saw. (Most blades are stamped "This Side Out" so that you will be sure to get it right.) Then tighten the flange and screw with the wrench.

To make a cut, set the depth and angle adjustments. Set the ripping guide if you are cutting a board lengthwise or cutting a panel with a straight edge. For crosscutting, mark a guideline on the board to be cut. Place the front of the shoe on the work so that the guide mark or notch on the shoe (indicating the location of the blade) and the guideline are aligned. With the blade clear of the work, start the saw and let it attain full speed. Advance the saw, keeping the guide mark and guideline aligned. Do not force the saw; let it cut at its own pace. If the saw stalls, back it out slightly, keeping your finger on the trigger. When it resumes cutting speed, advance the saw again. When the cut is completed, make sure the saw blade comes to a complete stop before setting the saw down.

Eternal vigilance should be the watchword when working with a circular saw. For safety's sake, keep your mind fixed on the job at hand. Make sure the work you are sawing is

The upper blade guard is fixed; the lower blade guard is movable.

The lower blade guard slides up and back as the blade cuts into work.

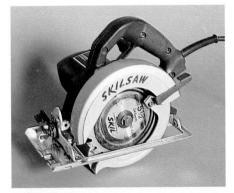

Cutting angle adjustment allows shoe to be tilted up to 45°.

The ripping guide extends from the shoe to guide the blade.

A knob or handle on the front of a circular saw makes possible two-hand operation, providing better control for guiding the saw along a cutting line.

Start the saw with the blade clear of the work, then advance it into the cut without forcing it which can cause the blade to bind.

Make sure work you are cutting is well supported. In this case, the saw cuts the saw horse slightly, but it can be protected with a piece of scrap.

properly secured so that it can't slip or turn; if you can't hold it firmly with your hand, secure it in a vise or clamp it to a bench or other support. Use only sharp blades, so that they do not stick in the work and kick back at you. Never start the saw with the blade against the work; this too can cause kickback, and it puts extra strain on the motor. Keep in mind that you can't see the blade. The part of the blade above the work is guarded, but the part that projects below the work is not. Those exposed teeth are extremely dangerous, so make sure that no part of your body gets in their way. Always keep a firm grip on the saw's handle, and never set the saw down while the blade is still turning. A healthy respect for the circular saw is your best assurance of safety in its use.

Portable jig saws

The portable *jig saw*, also called a *saber saw*, can perform a variety of cutting jobs, depending on the blade used with it. It will crosscut or rip boards, although not as quickly or accurately as a circular saw. It will cut curved shapes and scrollwork, and it will make its own starting hole for plunge cuts, such as openings in wall paneling for electrical outlets, or in countertops for sinks.

The maximum depth of cut of a low-cost, light-duty jig saw is about 1½ inches in soft wood and 1 inch in hard wood. If you try to cut greater thicknesses, the saw will stall or overheat. Heavy-duty models using a 6-inch blade can cut depths up to 3½ inches of soft wood and 2½ inches of hard wood.

If you intend to cut only simple shapes with a jig saw, a one-speed model will be satisfactory. For more complex work, a two-speed or variable-speed model will give you greater control. Of course, it will also cost more.

The shoe of a jig saw is slotted at the front to accommodate the blade. On all but the least expensive models, the shoe also includes a cutting-angle adjustment and a place to attach ripping or circle guide accessories.

The portable jig saw makes cutting irregular shapes an easy job. It can be maneuvered along almost any pattern line you draw.

Here, a ripping fence guides the portable jig saw for a cut parallel to the edge of the work far more accurately than a freehand cut.

By tacking the end plate of the ripping fence in the center, the portable jig saw becomes a perfect circle cutting tool.

The ripping fence, for guiding the saw blade parallel to the edge of the work, can be quickly attached, adjusted, or removed by turning screws. Most designs can also double as circle-cutting guides; the end plate of the fence is tacked to the center of the circle, and the rod adjusted to the proper radius.

The cutting-angle adjustment is a large, scaled hinge between the shoe and the housing; it allows precise cuts at angles from 90° to 45°. The hinge is loosened by a lever, wing nut, screw, or hexagonal key. Then the shoe is tilted to the desired angle and the hinge is tightened.

Some jig saws have a palm grip on the housing, but most are fitted with a handle on the top; others may have auxiliary bars or knobs to help keep the saw steady or to guide the blade. A slide, toggle, or trigger switch turns the saw on and off. The first two are self-locking for continuous operation; a trigger switch must be locked in the "on" position by pushing a button, and it shuts off instantly when you squeeze the trigger. For a variable-speed jig saw, a trigger switch that controls blade speed in response to finger pressure is convenient. Variable-speed models with slide or toggle switches have separate speed controls.

This portable jig saw has an adjustable, orbital blade control to adjust pitch.

Rest the shoe on the work and lower the blade to make a plunge cut.

with only moderate skill can use this tool to produce artistic edgings, multi-curved moldings, relief panels, delicate grooves for inlay work, perfect rabbets and dovetails for joints, and precise mortises.

A router is simple in concept. It consists of a high-speed (usually 25,000 to 30,000 rpm) motor with a chuck on the end mounted on a flat base plate. An almost infinite variety of router bits can be chucked into the tool for grooving and shaping.

Some smaller routers are meant to be held with one hand, but most have handles or knobs on two sides and are meant for two-handed operation.

Keep a firm grip when using the router—its high speed tends to make it twist when entering the work. A router needs to be held down rather than pushed along. The operating speed pulls the bit through the work. With a little practice, you will get the feel of the router, enabling you to take advantage of its potential. Just remember to keep fingers, hands, sleeves, and everything else except the work away from that bit. It spins at an incredible 500 turns per second —enough to chop through just about anything that gets in its way.

Accessories for the router include straight, circle, and slot guides, as

There are several types and sizes of blades for use in jig saws. Woodworking blades include coarse-toothed, for fast, rough cutting; fine-toothed, for smooth, finish cutting; and thin, narrow types for scroll cutting. Most hardware stores sell general-purpose assortments at reasonable prices. Blades are easily changed; the screws that hold them in place are loosened or tightened by either a screwdriver or a small hexagonal key.

To cut from an edge with a jig saw, allow the forward end of the shoe to rest on the work surface without making blade contact. Line up the blade with the cutting line, turn on the saw, and advance it into the work, keeping firm downward pressure on the saw but applying only slight forward pressure. Forcing the cut will not increase cutting speed, but it may cause blade damage.

To make a plunge cut (cutting directly into a surface without a starting hole), rest the saw on the front edge of the shoe without making blade contact with the work. Hold the saw firmly, start the motor, and gradually lower the blade into the work surface. When the shoe is in the normal position proceed with the cut.

Routers
The *router* has an almost limitless capacity for cutting intricate contours in wood, and even an amateur

The router is one of the most versatile tools you can have in a home workshop. Here, one quickly cuts a perfect dado groove.

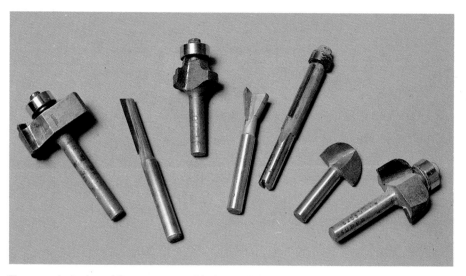

These are just a few of the many router blades available for making straight-line and decorative edge cuts for such things as frames or cabinet doors.

Some smaller routers can be operated with one hand when using both hands would be inconvenient. This model also has knobs for two-hand use.

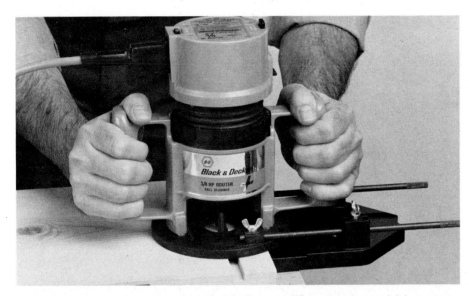

This accessory guides the router along straight-line cuts. When turned around, it becomes a guide for cutting accurate circles.

10,000 opm, but stroke length is halved. The sanding capacities of the two arrangements are similar, but shorter strokes produce a glossier surface. Reciprocal sanders with vibratory motors have a speed of 7200 strokes per minute (spm), equivalent to 7200 opm, and a short stroke. They produce a fine finish but smooth more slowly than orbital models.

The size of the sanding pad (platen) helps determine the work capacity of a sander. These range from 3½ × 7 inches to 4½ × 9 inches. Manufacturer's catalogs and brochures often list only the size of the abrasive sheet required to fit the pad (usually one third or half of a standard sheet). To determine pad size, subtract 2 inches from the given sheet length.

The method of changing abrasive sheets depends on the model. Some have a lever to open and close clamps at each end of the platen. On others, a special key or a screwdriver is required to loosen and tighten platen clamps. Still others have spring-loaded clamps that must be held open while an abrasive sheet is inserted.

Most finishing sanders have separate handles, but some simply have a palm grip on the housing. Try such a model under power before purchasing it to find if it becomes uncomfortably hot during extended operation.

well as special template guides that can be pre-cut so that you can rout just about any shape. Special tables hold the router upside-down, with the bit up, so that it can be used as a stationary shaper. Such tables should be equipped with guards to protect the user. These guards must never be removed if the tool is to be safely operated.

Finishing sanders

The *finishing sander* is the most popular tool for smoothing wood surfaces to which a finish coat is to be applied. With the proper abrasive, it can also be used for rougher work (see *Sanding and Abrasives*).

The least expensive finishing sanders usually have vibratory motors and back-and-forth, or reciprocal, sanding motion. Slightly more expensive models with rotary motors have a circular, or orbital, motion. Dual-action models can have the motion changed by movement of a lever, from orbital action for fast removal of material to reciprocal, straight-line sanding for a "hand-sanded" appearance.

Sander speeds are coordinated with the lengths of sanding strokes made by the machine. On some orbital models, the speed is between 3000 and 4500 orbits per minute (opm); on others, the speed is around

A finishing sander may have either orbital or reciprocal motion.

Heavier units may have front knobs or handles to provide better control and more uniform pressure with two-hand use.

The sander should be turned on before the tool is applied to the work surface. This prevents rough action that would occur when starting under load. Lower the sander to the work and guide it across the surface in long, slow strokes, applying uniform but slight pressure. Let the weight of the sander do the work. Excess pressure on the tool slows the sanding action and causes the abrasive paper to wear prematurely. Always keep the sander moving while it is on the work surface. Stopping in one place, even for an instant, will cause gouging and unevenness.

Belt sanders

A *belt sander* quickly sands large surfaces such as floors, planks, and paneling. It can remove old paint, erase scratches, and smooth uneven workmanship, but it is not suited for use on most furniture or for fine finishing.

Belt sanders cost considerably more than finishing sanders, and are literally heavyweights (6 to 20 pounds). Heavier sanders usually have greater work capacity, but they also take some muscle to control, especially when used vertically, as in sanding wall paneling or doors.

Other factors affecting the work capacity of this sander are belt size and speed. These are noted either on the tool label or in the manufacturer's literature. Belt size is expressed (in inches) as the width and circumference of the abrasive belt that fits the sander. Common sizes are 3 × 18, 3 × 21, 3 × 24, 4 × 21, and 4 × 24 inches. Generally, the larger the belt, the greater the work capacity. Belt speed is expressed in surface feet per minute (sfpm), and ranges from 900 to 1600. The higher the sfpm, the greater the work capacity of the tool for its size, or the faster it removes material from the surface being sanded.

The belt sander housing covers the motor, the upper surface of the abrasive belt, and one side of the sander. The other side is open to facilitate belt changing. Rear handles are usually a contoured D shape

The palm-grip finishing sander is for one-hand use.

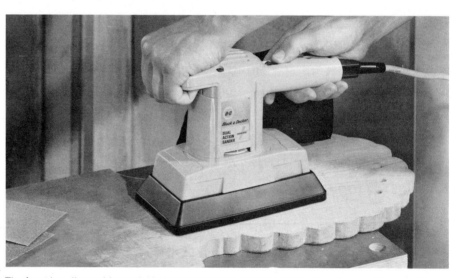

The front handle on this model helps to maintain uniform pressure on the work to avoid uneven sanding.

An attached dust bag for collecting waste material is a useful accessory for a belt sander.

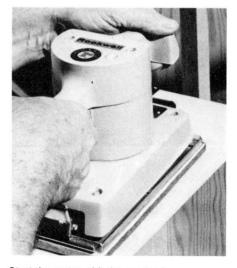

Start the motor with the sander free.

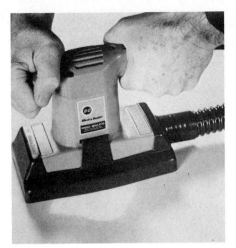

Some finishing sanders can be fitted with flexible hoses for carrying dust away.

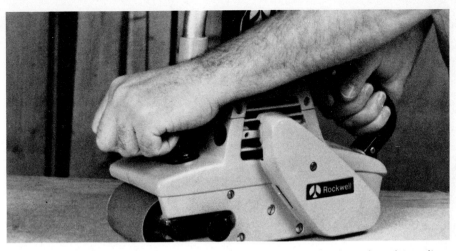

To remove the maximum material, hold the sander diagonally across the grain and move it with the grain.

to provide a good grip. A knob or bar handle is helpful for guiding the tool and applying uniform pressure.

On some models, the rear roller, powered by the motor, has a soft outer layer that grips the abrasive belt to apply traction and prevent slipping. Other models have a soft lining belt over which the abrasive belt is placed for good traction. To change a belt, the free-running front roller is retracted and the old belt removed. After the new belt is on and the front roller returned to operating position, the belt must be aligned so that it won't run off to one side during

sanding. The owner's manual will contain instructions for belt changing and alignment.

A belt sander produces a large quantity of waste (dust) from a work surface. Some form of dust collector is very desirable. Some models have built-in dust bags; for others, dust bags can be bought separately and attached when needed. On still others, a flexible hose can be attached. One end is connected to a vacuum cleaner, which is turned on to suck dust away from the sander.

Always wear goggles when you operate a belt sander—even a

sander with a dust collector. As with a finishing sander, the switch (a trigger with a locking button) should be turned on before the belt is in contact with the work. When the tool is at operating speed, lower the belt flat on the surface. Never tilt the tool or hold it in one spot—this will damage the work. For removing a lot of material, hold the sander diagonally across the grain, but move it with the grain. Apply light, uniform pressure. For final smoothing, the sander should travel parallel to the grain of the wood.

4 STATIONARY POWER TOOLS

When you advance to the big leagues of woodworking and begin doing a lot of fairly complex projects you will want to have some heavy hitters—stationary power tools that can turn out work far more quickly and accurately than any hand-held tool. Of course, you pay for their capabilities. A fully equipped home workshop can represent an investment of several thousand dollars, so the casual or occasional woodworker may settle for one or two major tools (a bench saw or radial arm saw, perhaps a drill press) or even resist buying any rather than tying up money in equipment that may be used only once or twice a year. But if you are bitten by the woodworking bug, you will eventually want, in addition to that saw and drill press, a full complement of workshop tools: band saw, jig saw, belt/disc sander, shaper, jointer, and lathe. With this lineup and your mastery of their use, you can do just about anything with wood except turn it back into a tree. But be careful: all of the safety cautions discussed in the previous chapters apply doubly to stationary power tools.

Shopping for stationary power tools

Be a careful comparison shopper when you are in the market for a stationary power tool, for all are not created equal. A bench saw is a bench saw, but one saw may incorporate features and accessories that gives it a greater range of capabilities than others. Peruse the manufacturer's lit-erature before you buy, and ask a salesman to display the various features of the tool. You can then see how it stacks up against others of its type. Often, when craftspeople are satisfied with a tool, they will outfit their entire workshops with tools of the same brand. That's not necessarily a bad idea, but you should compare just the same.

Almost as valuable as the tool itself is the instruction book that comes with it. It tells you exactly how to use that particular tool and all its accessories, and how to operate it safely. Study the book carefully and always follow its recommendations. Keep the book near the tool for ready reference when necessary.

Bench saw

The *bench saw* (or table saw or stationary circular saw) is probably the most used tool in a home workshop (except in the workshops of those who prefer a radial arm saw). It can crosscut, rip, miter, bevel, and, with the proper accessories, cut dadoes, make moldings, and even sand stock. Various guides help it perform with a high degree of precision.

Many older bench saws were made with tilting tables; that is, the saw table was adjusted to make angle cuts while the blade remained vertical. With this arrangement, the work had to be held at an angle while being fed into the saw—often an awkward and inconvenient proposition. Today, almost all bench saws have tilting arbors. The table remains horizontal, and the blade is adjusted for cuts from 90° to 45°.

Size designations of bench saws are determined by the largest blade the tool will accommodate. An 8-inch saw is the minimum choice for most home workshops. It has a depth of cut of about $1\frac{3}{4}$ inches at 90° and $1\frac{3}{8}$ inches at 45°. Also available are 9-, 10-, and 12-inch models, with the latter having depths of cut of about 4 inches at 90° and $3\frac{1}{4}$ inches at 45°. This may be more capacity than you will ever use, so consider the type of work you will likely be doing before deciding on the size bench saw you will buy. Remember, the bigger the saw, the higher the price.

Blade height is adjustable and should be set so that the gullets between the teeth just clear the surface

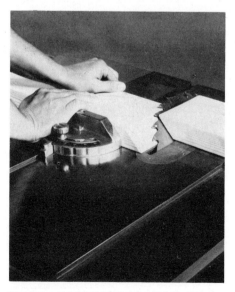

Cranks on the front and side of saw control table tilt and blade height.

of the work. The blade rotates toward the end of the table from which the operator feeds the work, helping hold the work to the table and control the cut. A miter gauge, adjustable from 90° to 45°, fits into a groove in the table top to guide crosscutting operations. A ripping fence along the length of the table can be adjusted to the desired board width. For bevel cutting, the table is tilted, usually by means of a crank, to the desired angle.

The bench saw should be disconnected from the power outlet—not just turned off—while adjustments are made. When the saw is started, allow it to reach full operating speed before feeding in work. Do not force work into the blade; let the saw do

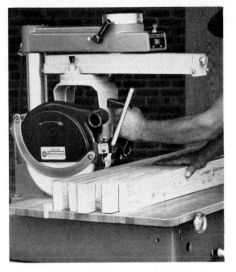

For cross cuts, radial arm saw blade is pulled across the work.

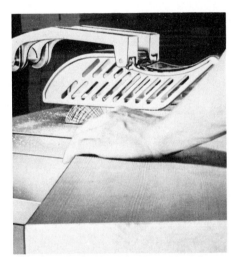

Blade guard is left in place for most cuts, riding over the work.

the cutting at its own rate. Keep the work snug against the miter gauge or ripping fence, and stand to one side of the blade—never in front of it. Needless to say, keep your hands away from the blade! A blade guard covers the blade for most operations. Keep it in place whenever possible. The guard will have to be removed for some jobs, but put it back as soon as possible lest it be forgotten the next time the saw is used.

Radial arm saw
The *radial arm saw* performs basic sawing operations—crosscutting, ripping, mitering, beveling—along with more complicated jobs such as

compound angle cutting. With accessories and attachments, it can be adapted to dadoing, shaping, molding, sanding, drilling, boring, and routing—all with speed and great accuracy.

The radial arm saw differs from a bench saw in that the blade cuts into the work from above. For many operations, this gives the operator a better view of what is being done. Another difference is that for most operations the work is stationary—the saw, mounted on an overhead arm, is moved through the work.

The saw head and blade can be swung 360° on most models and can be tilted for angle and compound cuts. A 10-inch saw typically has a maximum depth of cut of 3 inches at 90° and 2¼ inches at 45°. A 12-inch model can cut stock up to 4 inches or more. Cutting depth is adjusted by means of a crank.

For crosscutting and mitering, the work may be clamped to the table or held against the fence at the rear. For ripping, the saw head is set and locked at a 90° angle to the arm. The work is then fed into the blade from the direction marked on the blade guard.

Most radial arm saws include several safety features. One is an anti-kickback rod on the front of the blade guard. This should be adjusted so that it lightly touches the top of the work. Another safety device is a lock-

able power switch to prevent accidental starting. Always follow the manufacturer's safety recommendations.

Band saw
The *band saw* consists of two large pulleys over which is looped a continuous, flexible steel blade. It can be used for crosscutting and ripping, but the band saw is best suited for cutting curves and fine scrollwork. Most band saws have tilting tables to allow work to be cut at an angle.

Size is stated as the diameter of the pulleys. The width of work that can be cut is limited to the distance between the blade and the frame (usually ¼ to ⅜ inch less than pulley size). For example, a 10-inch band saw may handle work up to 9⅝ inches wide; a 14-inch saw unit may take work up to 13¾ inches wide. Most band saws can cut wood up to 6 or 6¼ inches thick.

Band saw blades for cutting wood come in a variety of widths, from ⅛ to ¾ inch. Points per inch range from about 5 to 15. The thinner blades and finer teeth are for cutting a sharp curve and tight radius; the larger ones, for straight cutting and a slight curve. As a rule of thumb, use the largest size blade that will do the job.

Band saw accessories include miter gauges and ripping fences, useful for crosscutting and ripping operations. However, the band saw will not

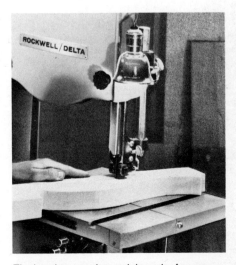

The band saw makes quick work of curves and irregular shapes.

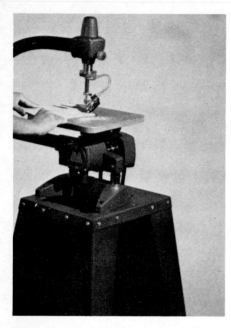

The jig saw cuts intricate curves and other shapes, as well as inside openings in wood or panels.

make these cuts as accurately as either a bench saw or a radial arm saw.

Jig saw

A *jig saw*, or *scroll saw*, is used for light-duty cutting of straight lines and intricate curves. Its blade is fastened at top and bottom and moves up and down. Its size is determined by its cutting clearance (the distance between the blade and the frame). Popular sizes range from 18 to 24 inches.

A unique feature of the jig saw is its ability to make interior cuts. The blade is removed from the saw first, and a pilot hole is drilled in the work. The blade is then passed through the hole and retightened in the saw.

Many jig saws have tilting tables so that angle cuts can be made. The maximum depth of cut on most models is about 2 inches. A hold-down device is located along the blade. This should be adjusted so that it just touches the surface of the work in order to offset the vibrations caused by the reciprocating action of the blade.

Jig saw blades come in many sizes. The narrower and finer the blade, the tighter the radius it cuts.

Drill press

There are many tools for making holes in wood, but for precision drilling, angled holes, or repetitive drilling of holes to a specified depth, there is only one tool: the *drill press*. With the proper fittings and accessories, a drill press can make square holes and mortises; it can also plane, shape, rout, and sand wood.

A drill press has four basic parts: base, column, table, and head. The table is adjustable and holds the work. (On some models, the table also tilts.) The head contains the motor, chuck, feed arm, and controls. The head tilts up to 90° for angle drilling and swivels a full 360° around the column, offering almost unlimited flexibility.

Drill press capacity is measured by doubling the horizontal distance from the center of the chuck to the supporting column. This is the maximum width of stock in which the tool can drill dead center. For example, an 11-inch drill press measures 5½ inches from chuck to column and can drill in the center of work 11 inches wide. Other popular sizes are 15-, 16-, and 32-inch. Chuck capacities on most models are ½ inch. Twist bits (the same kind used in portable electric drills) are used for making holes with diameters up to ½ inch; spade bits, for ⅜ to 1 inch. For larger holes, there are hole saw attachments. Other accessories include fly cutters (for cutting discs) and plug cutters (for making plugs of exact sizes). A mortising attachment —a hollow chisel and bit—is used to cut square holes.

To use the drill press, raise the table so that the work is as close to the bit as the job allows. If the piece is to be drilled through, place a block of scrap beneath it so that the bit doesn't splinter the wood as it emerges. Hold the work firmly or clamp it to the table. Start the machine and let it attain full speed before lowering the drill into the work.

Shaper

The *shaper* is a high-speed tool for, obviously, cutting various shapes in wood; you are limited in choice of shape only by the number of cutters or bits you have. The shaper can make moldings, decorative beads on panel edges, and flutings; with it, you can add a professional finishing touch to any do-it-yourself project.

Because of the speed at which the shaper operates, it commands special respect. Always make sure that

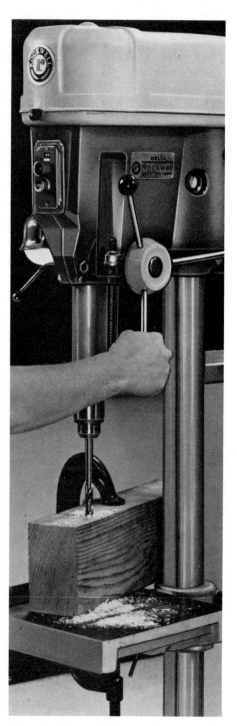

Work should be held firmly or clamped to the drill press table.

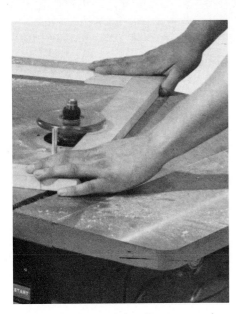

With the fence removed, a shaper can make cuts on inside edges.

the cutters are sharp and fastened securely in the tool before starting the motor. Keep safety guards in place at all times. For straight cutting, use the adjustable fence on the shaper table. For irregular edges, remove the fence and hold the work against the spindle collar. Feed the work slowly, at an even pressure and speed, and keep your hands well clear of the cutter. Always wear safety goggles when working with the shaper.

Jointer

Although the *jointer* does a limited number of jobs, it does them extremely well, and fine cabinetry would be next to impossible without this tool. Most readily available commercial lumber, although dressed, is rough-cut and imperfect. The jointer is used to finish the lumber surface prior to cutting on a bench or radial arm saw. The jointer is so named because it planes wood so flat and smooth that when pieces are joined together with glue they fit perfectly.

The essential parts of a jointer are the base, front and rear tables, cutter head (with three or four blades), safety guard, and fence. The size of the jointer is determined by the length of the knives, which also determines the widest boards it can cut. Common sizes for home work-

shop use are 4-, 6-, and 8-inch. The front table (and, on some models, the rear table as well) is adjustable to control the depth of cut. The fence tilts and locks for bevel cutting. The safety guard should be left in place at all times.

For safety, use push blocks when surfacing a board on the jointer. Hold the board firmly against the fence and the table, and do not stand directly behind the work. Always wear safety goggles when operating this tool.

Belt/disc sander

The combination *belt/disc sander* smooths regular or irregular surfaces and edges for finishing. The disc is used for fast removal of material and for finishing the ends of boards. For bevel sanding, the disc table tilts down to 45°. On some models, the disc is removable so that other accessories, such as drum sanders and buffing wheels, may be attached to the end of the drive shaft.

Final smoothing, as well as surface and edge sanding, is done on the belt. A backstop helps to control the work, but it can be removed for sanding longer pieces. Inside curves are sanded on the drum at the end of the belt. The belt tilts upward and may be locked at any angle from horizontal to vertical. Always wear safety goggles when using the sander.

Push blocks are not used on the jointer in edge planing when the operators hands stay safely above the cutter.

Faceplate turning is done at the motor end of the lathe.

Lathe

The *lathe* is a tool for the advanced and dedicated woodworker. With it, you can design and make such items as chair legs, decorative spindles, lamp stands, bowls, and trays. But the lathe demands more skill on the part of its operator than any other power tool, so if you decide to invest in one be prepared to devote plenty of practice time before you achieve perfection.

The size of a lathe is expressed as the distance between the bed of the tool and the center of the spindle that holds the wood for turning. A 10-inch lathe is a good choice for the home workshop. Another factor to consider is the length of wood that can be worked, usually 32 to 48 inches.

For spindle turning (as on chair legs), the work is mounted between a live center (turned by the motor) and a dead center (an unpowered bearing), and is rotated while the operator holds any of a variety of cutters against it. For faceplate turning (as when making bowls), the work is mounted on the powered end.

Follow all manufacturer's instructions and safety precautions when using the lathe. Never wear loose clothing (it might become caught in the work), and always wear safety goggles or a face shield.

One good way to tell woodworkers from wood butchers is to look at the joints in their work. How are the pieces of wood joined together? There are countless ways to do this, and countless variations on these ways. You will probably invent some of your own as you experiment with the basic joints and splices shown here. *Joints* are used to connect two pieces of wood that come together at an angle. *Splices* join two pieces of wood end-to-end on the same plane.

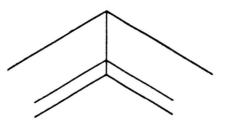

The butt joint: simplest joint in woodworking, and least strong.

Butt joints

The butt joint is the simplest of all joints. It is widely used in construction carpentry, but only rarely in fine woodworking or cabinetmaking. It depends on the fasteners (nails, screws, dowels) for strength, and so is relatively weak. When making this joint, check with a try square to be sure that the end of the piece to be butted against the surface of the other piece is perfectly square. Use the square to mark the surface to be joined so that the two pieces will be at right angles to one another.

Miter joints

Miter joints are used for picture frames, door moldings, and other applications where it is desirable to conceal the end grain of the wood. They are a form of butt joint, with the angle at the corner halved between the two pieces being joined. The most common miter joints are cut at

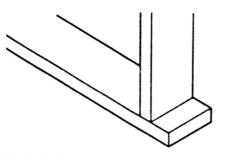

The miter joint: most commonly used for frames and molding projects.

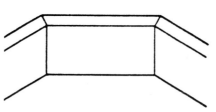

Hexagonal project with miter joints: not all miters are 45°.

an angle of 45° and joined for a 90° corner, but the angle can be varied to suit the work at hand—for example, 60° cuts for a hexagonal project.

For accuracy, miter cuts are best made with a power saw adjusted to

The radial arm saw can be set to cut precise miters at any angle.

the desired angle, or in a miter box with a back saw or other fine-tooth saw. Use a square to check the fit of the pieces to be joined before assembling.

How to cut a miter in a miter box

Cutting miters is easy, but practice this sequence a few times before trying to finish work.

STEP 1
Selecting the angles
Set the miter box at the angle at which you want to cut the wood. The instructions that come with the tool detail this step, and it is not difficult.

STEP 2
Cutting the first piece
Set the piece of molding that will

mate to an adjoining surface firmly in the box and then cut off an inch or two. The purpose here is just to get a clean, finished end. Take the piece of stock (you may have to cut it down into smaller pieces to facilitate working with it) and set it in position on the item on which you are working, at the point where you want the end of the molding to fall. Mark the shape of the adjoining face onto the stock. Position the stock in the box, so that when you make your second and final cut the saw will just follow and obliterate the drawn guideline as the saw meets the edge of the stock.

If you have cut the miter a little long, you can carefully trim it with a table saw. It is better to cut too long than too short; if you cut the stock too short you must start over again.

STEP 3
Cutting and attaching the mating pieces

Cut the other mating pieces in the same way. When all the pieces are cut, fit them together and nail them in place on your cabinet or bookcase— or glue and nail them together to make a picture frame. A good tool for nailing the pieces is the brad driver.

When the brads are in place, set them with a nailset. Fill the depressions with wood putty, then stain or paint to match the stock.

Lap joints

Lap joints find many uses in furniture construction, kitchen cabinets, and similar projects. Full laps are used when boards of different thicknesses are to be joined—a 1 × 4 to a 2 × 4, for example. The 2 × 4 is notched 3/4 inch (the thickness of the 1 × 4) to receive the thinner member. Half laps are generally used when joining two pieces of the same thickness. Each piece is notched half its thickness to make the joint.

The most common lap joints are the *end* or *corner lap*, the *cross lap*, and the *middle lap*. A variation is the *dovetail lap*, the joined pieces are "locked" together, resulting in a very strong joint.

Using an adjustable miter box.

Start with a clean cut at the end.

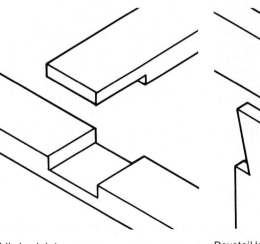
Put the cut end in place and mark the other end for the last cut.

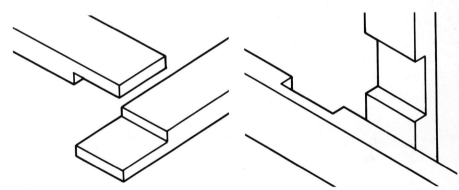
End lap joint (corner lap).

Cross lap joint.

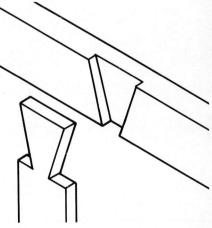
Middle lap joint.

Dovetail lap joint.

Making a corner half lap with a stationary power saw

Making a half lap with a power saw is an efficient operation. Be sure to keep your fingers clear of the blade when making the cut.

STEP 1
Making the first cut

Measure in from the end of each board by the width of the board to find and mark the overlap dimension. Draw lines across the widths of both boards at these markings. Then place the boards next to each other and set the saw so that it will cut across both lines at half the depth (thickness) of both pieces. Make the cuts.

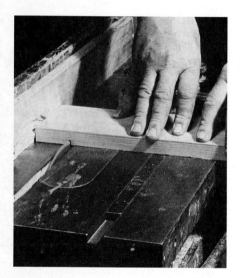

Put the pieces of stock side by side and cut to half lap depth.

Stand up stock and run it through the blade with a push board.

STEP 2
Making the second cut

Set the fence half the board thickness from the blade. Set the depth (thickness) of cut the same as the width of the board. Stand one board on end (turn it 90°, on edge) and, using a push board, push it through the saw. The excess piece will fall out. Repeat the procedure on the end of the second board.

STEP 3
Finishing the joint

It is possible that the fit will not be perfect—one board may overlap the other just a little. If so, use a sharp plane or sandpaper to shave off the excess. The joint is fastened with screws or nails, and usually glue, depending on the requirements of the particular piece of work.

Making a middle lap with a hand saw

Cutting a middle lap by hand is a little harder than using a machine.

STEP 1
Marking the pieces

Mark gauge lines on both pieces indicating the depth of the cuts to be made (half the thickness of the wood). Make sure the pieces are squared up, then carefully mark the width of the cross piece on the receiving piece.

STEP 2
Making the first cut

Clamp the receiving piece to hold it steady and make cuts at either end of the portion to be removed and one in the middle; work within the inner edges of the gauge marks. Chisel away waste wood, working from either end toward the middle.

STEP 3
Making the second cut

Clamp the cross piece and cut along the gauge line just to the mark indicating the required depth (called the *shoulder line*), again cutting just within the gauge line. Then cut across the shoulder line and remove

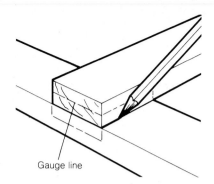

A gauge line indicates half the thickness of the end to be cut.

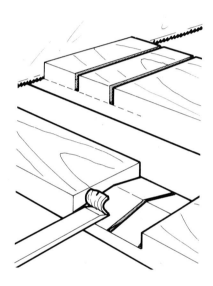

Chisel away the waste wood between the kerf cuts.

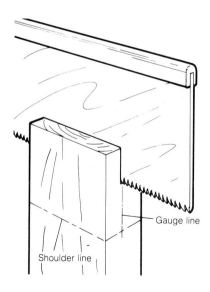

Cut along the gauge line down to the shoulder line on the crossrail.

waste wood. Finish the joint as described above.

Rabbets

Rabbet joints are frequently used to assemble corners in drawers, bookcases, cabinets, and similar projects. The joint is formed by cutting a recess (*rabbet*) in the end of one piece to accommodate the second piece to be joined.

The rabbet can be cut in several ways: with stationary power tools, with repeated passes of a circular saw, or with a router with an appropriate bit. Bits are commonly available, but the sizes you can make are limited, up to ½ inch wide. This cut can also be made in one pass with a dado head.

Outline joint on end of stock, then adjust fence and blade height.

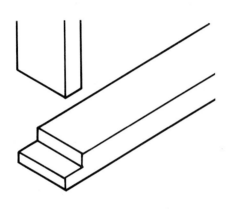

Rabbet joint: one piece to be joined is shaped, the other left plain.

Making a rabbet on a table saw

This is the quickest way to cut out a rabbet.

STEP 1
Making the first cut
Draw the outline of the piece to be cut out onto the end of the stock. Set the blade at the depth required (half the thickness of the stock is standard) and run the stock through.

STEP 2
Making the second cut
Turn the wood on its side to make the width cut. You can simply set the saw to the intended width, but it is also a good idea to mark the dimension along the top of the board so you can

Turn the stock and cut the other side for a square rabbet.

Two cuts complete the operations producing a clean rabbet.

be sure the cut is correct. Run the board through.

That is all there is to the rabbet—make two cuts and a neat rectangular piece of waste will drop out. To ensure that the edges of the cut are square, stick the mating piece of stock in place to see how it fits. If it does not fit well, then adjust the blade or stock position as necessary, and run the joint through again. Sand the cut.

Dadoes

Dadoes are channels cut across the grain of a board, into which a second piece of wood is fitted. They are widely used in the construction of tables, cabinets, and shelving. The most common is the housed dado, in which the entire width of the second ("housed") piece fits into the channel, or dado.

There are several variations of this joint. A *stopped* or *gain dado* does not extend across the full width of the board, and the housed member must be notched to fit. A *stopped housed dado* or combination dado and rabbet is cut across the full width of the board with a rabbet cut on the housed member. A *dovetail dado* has one edge of the channel cut at an angle, with the housed member notched at a matching angle and fitted into the dado from the edge of the board. This makes a very strong joint.

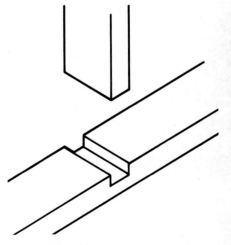

Housed dado joint: used in shelf construction. These are most easily made with a radial saw or router.

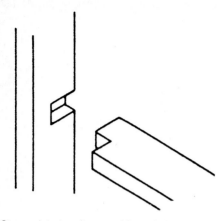

Stopped dado: often used for a dado's strength, with joint hidden.

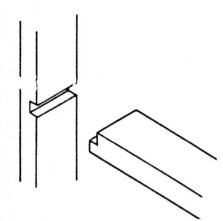

Stopped housed dado: a combination of dado and rabbet joints.

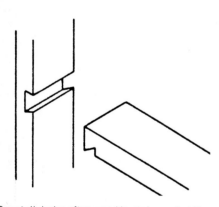

Dovetail dado: often used in stairs and at the backs of drawers.

Making dadoes several ways

The hardest method for making a dado is the chisel-handsaw-sweat method. It is far easier to use a table saw, a radial arm saw, or a router.

If you have a table saw, there are several ways to proceed. First, consider dado blades. There are two round blades, as well as knifelike rakers. Each has a certain thickness—usually ¼ inch for the round blades and ⅛ inch and 1/16 inch for the rakers. To achieve the desired dado width, you use as many round and raker blades as required in order to add up to the width size you require; then set the depth that you wish.

To prevent expensive mistakes when learning to work this joint—or any joint, for that matter—experiment with scrap wood until you achieve the correct cut.

Another useful tool accessory when making dadoes is the dado head. This looks like a miniature saw blade. It comes with an attachment that allows you to install it on a standard table saw or radial arm saw: the dado head is twisted; this design provides a cut that is wider than the blade edge. Full instructions for its use come with the device. Again, it is important to practice on scrap until you have the head set to the exact width and depth that you will need.

The easiest method to cut a dado is with a router. There are many sorts of router bits available to cut whatever width and depth you wish. Before you begin to cut the dado, use C clamps to secure a straightedge in line with the cut you want to make. The straightedge will act as a guide for the baseplate of the router. Clamp the straightedge so that the cutter will make the dado in the exact spot you wish. You can use a ruler and square to measure and mark dimensions, or you can work it out first on scrap.

No matter what tool you use to make the groove for the dado, make sure the cut is smooth and the edges sharp so that it can readily accept the mating piece of wood. A piece of tightly folded sandpaper usually works to clean the cut after it has been made. If the cut is very rough, go over it again with the router.

Mortise-and-tenon joints

The mortise-and-tenon is a strong joint used in the construction of

A radial arm saw with a dado blade cuts a dado groove in one pass.

The dado head is formed (twisted) to make a cut wider than blade thickness.

This dado set makes cutting dadoes easy on a table saw. The raker blades (center) are fixed between the round blades (either side).

desks, tables, chairs, and cabinetry that will be subjected to heavy wear. A *mortise* is a rectangular cavity cut into a piece of wood, into which a *tenon*—a projection cut on the end of a second piece—is inserted. The strength of the joint depends on the accurate fitting of tenon to mortise.

A *blind mortise* extends only part way (usually one half to three quarters) through the work; a *through mortise* goes through the board. The tenon for a blind mortise should be cut 1/16 inch shorter than the depth of the mortise; this allows space for glue. The tenon for a through mortise may be cut slightly longer than the thickness of the mortised piece, then trimmed off flush after assembly.

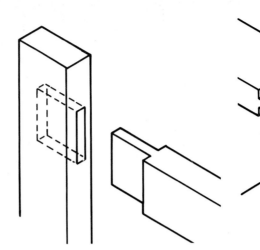

Blind mortise-and-tenon joint.

Through mortise-and-tenon joint.

Making a mortise-and-tenon joint

This procedure takes some practice to perfect. Try these steps with pieces of scrap before you move on to finish work.

Routing a dado groove with a guide.

Sharpening the groove with sandpaper.

STEP 1
Cutting the tenon
Lay out the tenon by measuring from the end of the work the length of the tenon. Use a try square to mark this line on all four sides of the piece. For general purposes, the tenon should be one third to one half the thickness of the board on which it is being cut. Locate the exact center along one edge and measure half of the tenon thickness on each side of this center point. Use a try square to extend these markings the length of the tenon, across the end, and down the other side to the previously marked shoulder lines indicating the end of the tenon. Follow the same steps to mark the width of the tenon on the stock (usually a cut of at least 3/4 inch at the top and 1/4 inch at the bottom). With a back saw, cut along these guidelines to the depth of the shoulder lines. Cut the shoulders to the guidelines.

STEP 2
Marking the mortise
The length of the mortise must equal the width of the tenon. With the try square and a sharp pencil, mark lines across the work to indicate mortise length. Locate the exact center of the mortise and measure out from the center exactly half the width of the mortise (which corresponds to the thickness of the tenon). Outline the mortise, then check your markings by placing the tenon on the surface

of the work. Score the outline with a sharp knife.

STEP 3
Drilling out the mortise
Select an expansive or auger bit slightly smaller than the width of the mortise. Place the spur of the bit on the center line of the mortise and bore a series of overlapping holes, with the first hole just touching one end of the mortise and the last hole just touching the other end. Be sure to hold the bit perfectly perpendicular to the work during this operation (a drill press makes the operation easy). The holes should be bored 1/16 inch deeper than the length of the tenon. (For a through mortise-and-tenon joint, the holes are bored completely through the work. Place a piece of scrap beneath the work so that the wood does not splinter as the bit breaks through.)

STEP 4
Cleaning out the mortise
With a sharp chisel, pare the walls of the mortise to the guidelines, keeping them perpendicular to the surface of the work. Square the ends of the mortise and remove waste from the bottom with a chisel slightly narrower than the width of the cavity.

An *open mortise-and-tenon*, sometimes used in corners, is similar to a through mortise-and-tenon with one end of the mortise removed. The tenon is usually cut only on the

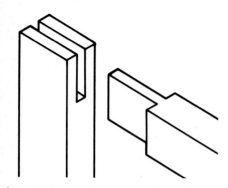

Open mortise-and-tenon joint; also called a slip joint.

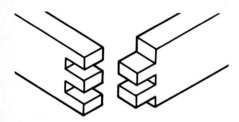

Pin joint also called a box joint, like a dovetail but with square pins.

sides, or on the sides and one edge, of the work. The mortise sides are sawed, as with a tenon, and the waste is removed by boring a hole near the inner end and trimming with a sharp chisel.

A *pin joint*, used primarily in box construction, is formed by making a series of open mortise-and-tenons in line. This joint is also called a finger lap.

Dovetail joints

Dovetail joints are used for fine furniture, drawers, and projects where good appearance and strength are desired. The strength of this joint is derived from the flare of the projections (pins) on the end of the board, which fit exactly into sockets between dovetails on the mating board. The strongest joints are those in which the pins and the dovetails are the same size, but for the sake of appearance, the dovetails are usually wider than the matching pins. They should not be more than four times wider, however, since this could cause the pins to break.

Single dovetails

Dovetail joints are a test of a woodworker's skills, so don't expect perfection on the first try. Practice making a through single dovetail.

STEP 1
Cutting the socket

Lay out the socket to the desired angle, then make the side cuts with a back saw and clean out the waste with a sharp chisel.

STEP 2
Cutting the pin

Place the socket on the matching piece and mark the outline with an awl or a sharp pencil. Then cut the pin with the saw and chisel.

STEP 3
Adjusting the fit

Check the fit. If it's too tight, use a sharp knife to pare off a thin sliver of wood from the pin or the socket side. If the fit is too loose or uneven—well, you'll have to practice some more.

Multiple dovetails

Once you have mastered the single dovetail, there are many variations using the same techniques. A *through multiple dovetail* joint is a series of single dovetails extending along the entire length of the ends of the joined pieces. After the first cut is made, a wood template can be made to lay out the remaining cuts. The *lap* or *half-blind dovetail* is often used for joining drawer sides to the front. The

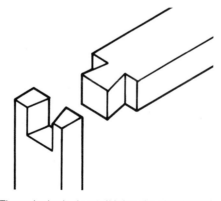

Through single dovetail joint: the simplest of the dovetails.

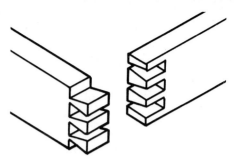

Through multiple dovetail joint: often used in drawer construction.

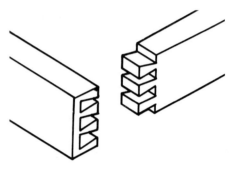

Lap dovetail joint, also called a half blind dovetail: a partly hidden joint.

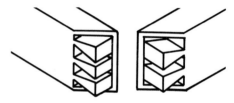

Blind dovetail joint, also called a blind miter or secret dovetail.

sockets on the front piece are cut only part way through, concealing the ends of the pins. The *blind dovetail* (also called blind miter or secret dovetail) has the sockets on both joining pieces cut only part way through, making an almost invisible joint.

Although the techniques for making dovetail joints are not difficult to master, they take study and practice. Different kinds of wood and different thicknesses of the pieces require different angles for the pins and sockets. When you have perfected the single dovetail, move on to an advanced woodworking book to learn the more complicated dovetails.

Splices

There are many ways to join pieces end-to-end. The *end butt* joins two squared ends, usually with glue. The *half lap* is like other lap joints, except that it joins pieces in a straight line. A *scarf joint* is made by cutting the ends of the two pieces at corresponding angles. A *face dovetail* applies the dovetail principle to in-line joinery. A *finger splice* requires very precise cutting; this splice should be attempted only by the advanced woodworker.

When structural strength is re-quired of a splice, it must be rein-forced. Metal fishplates or wood scabs are fastened on each side of the splice, preferably with bolts. Dowel reinforcement is adequate for lighter loads.

Dowel splices and joints

Dowels are often used to strengthen butt joints and miter joints and as a substitute for a mortise-and-tenon. *Dowels* are cylindrical pieces of wood, usually maple or birch, in diameters from 1/8 inch to 1 inch; they come in lengths up to 3 feet.

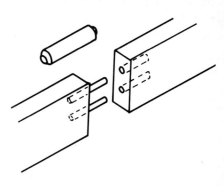

Dowel-reinforced butt splice: note chamfered ends on dowel.

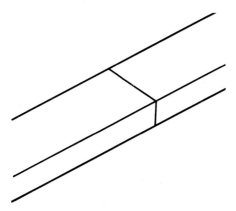

End butt splice: the simplest way to splice two boards, and the weakest; can be doweled for strength.

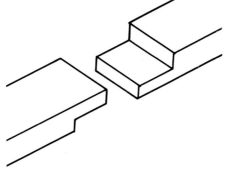

Half-lap splice: for more strength.

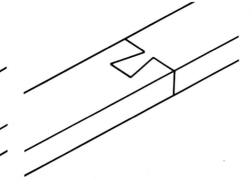

Face dovetail: locks boards tight.

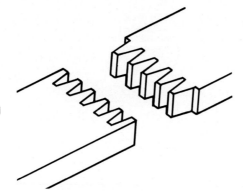

Scarf splice: stronger than a butt splice. Boards are beveled to fit together then nailed or doweled.

Finger splice: good for glue.

Doweled edge joint used to make up stock wider than commonly available.

Holes for dowel joints must be perfectly aligned for the joint to be flush, and they must be drilled verti-cally or the joint will not be true—actually, it might end up weaker than if dowels were not used at all.

With the aid of a doweling jig, most people can make precise, professional-looking joints and splices. There is really no good sub-stitute for a doweling jig. It makes the job simple and easy. You can use dowel centers and other ways of lin-ing up the dowel holes, but no other method or tool does the task right every time.

The fundamental purpose of the doweling jig is to force you to drill the two needed holes in exactly the right places—precisely opposite each other—so that when you insert the dowels the joined pieces fit exactly as you wanted them. You position the drill guide by measurement, so be careful. If your measurements aren't precise, the jig can't help.

Making a doweled joint

This is the most basic way to make a doweled joint. It is ideal for joining two pieces of stock where you need a width greater than you can find in the lumberyard.

STEP 1
Aligning the pieces

Clamp the two pieces together, as shown in the accompanying photographs. Use a pencil and a rule to draw a straight line across both

pieces, at the point where they are to be doweled. Now take the clamp off.

STEP 2
Aligning the jig

Put the doweling jig on the first piece, clamping the jig in place after sighting the penciled line through it. There is a graduated marking on the jig to help you with this positioning.

STEP 3
Drilling the hole

Insert your drill bit into the drill guide on the jig and drill a hole that is slightly deeper than half the length of the dowel. Repeat the process on the

other piece of wood. Most doweled joints have two or three dowels, not just one, so do the same task as many times as necessary.

STEP 4
Inserting the dowels

Coat each dowel with glue and insert the dowels into the holes in one of the pieces. If your joint has three dowels, insert all three into the same side.

STEP 5
Finishing the splice

Fit the other piece over the protruding glue-coated dowels, and use a

To use a doweling jig, align the pieces to be doweled and mark across both pieces at once with a square.

Clamp one piece of the work and fit doweling jig over the mark.

Drill through the jig to make the dowel hole.

If the doweling jig is used properly, all the holes will line up with precision.

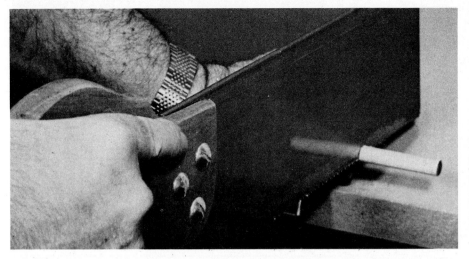

To make your own dowels, use hardwood doweling of the diameter you need, hand cut to the length desired.

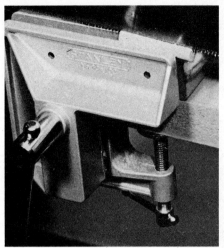

Clamp dowel pieces in a vise and cut in a channel to allow glue to escape.

rubber mallet to tap the piece onto the dowels until the joint is tight. Wipe off any excess glue around the joint. Then apply a clamp or clamps to hold the joint tight until the glue dries.

Surface doweling

Surface doweling is an easy method of reinforcing butt and miter joints, but one end of the dowel will be visible. Clamp the pieces to be joined together, then bore a hole through one into the other. Remove the clamp, apply glue to the mating surfaces, then reclamp—make sure the holes are perfectly aligned. Cut a dowel slightly longer than the depth of the holes. Chamfer one end, dip the dowel into glue, and drive it into the hole with a mallet. Wipe away excess glue, then cut off the protruding section of dowel with a sharp chisel.

Making dowels

You can make your own dowels by buying a hardwood dowel rod of the right diameter at your home center and cutting off dowels of the needed length. However, a dowel fits very snugly into the hole you drilled for it. So snugly, in fact, that as the dowel goes in, it may force glue to fill the bottom of the hole. The resulting hydraulic pressure can prevent complete entry of the dowel.

To remedy this, put your dowel in a vise and use the back saw to cut one or two slots, $1/16$ inch deep, in the dowel's side. The slots allow the excess glue to exit from the bottom of the hole as the dowel goes in. Finish the dowel by sanding the ends to round them slightly for easier entry.

Plywood joints

Plywood construction joints may be formed in several ways. A simple butt joint is suitable for $3/4$-inch plywood panels. For thinner panels, a reinforcing block in the inside corner

A butt joint is suitable for $3/4$-inch plywood (left); for thinner pieces, a wood corner block (right) is recommended.

Building a wood frame for support allows you to use thinner pieces of plywood for a job.

will make a stronger joint. Building a wood frame allows the use of thinner plywood.

Rabbet joints are often used for drawers and chests, but they should not be cut deeper than two or three plies. The same is true of dadoes, which are employed to make strong shelves with relatively thin panels.

Miters are an effective corner treatment, concealing the plies of both pieces. All except the butt joints should be cut with power tools for precise fitting.

Here, a rabbet joint adds strength to a corner made of two pieces of ³/₄-inch plywood.

Dadoes cut into plywood uprights make sturdy shelf supports without less sightly cleats to hold up shelves.

6 FASTENERS AND ADHESIVES

itting pieces of wood together is one thing; holding them together is quite another. With very few exceptions (some dadoes and dovetails that are self-locking), joints must be secured with *fasteners*—nails, screws, or other mechanical devices —or with *adhesives*.

Nails

Nails are the most primitive and least sophisticated fastening device. They are also the most widely used, especially in structural carpentry, because they can be quickly driven into place to form a sturdy joint. The holding power of a nail is achieved by the pressure of wood fibers, displaced by the nail entering the wood, trying to return to their original positions.

Nails have evolved considerably over the years. In colonial days nails were handmade, cut from iron sheets in a tapered shape with a blunt point and a rectangular cross-section. At about the time of the American Revolution, machines were developed to make cut nails. This type of nail is still manufactured for certain uses. Specially hardened steel cut nails are used for nailing into masonry— for example, attaching furring strips to panel a basement wall.

About a century ago the technology for making wire nails was brought here from Europe, and cut nail usage declined. The wire nail has a round cross-section with a head at one end and a point at the other; it is not tapered. There are many sizes and shapes, each designed for a particular job.

A *common nail* has a large, flat head and is used for most rough work. A *box nail* is thinner and is also used for rough work. A *finishing nail* has a small head—only slightly larger than the shank of the nail— with a depression so that a nailset can be used to conceal the head below the surface. A *casing nail* is similar, but the head is tapered and has no depression; it is often used for exterior trim work and may be driven either flush with the surface or set below it. *Brads* are small, lightweight nails with practically no head at all. They are used for light finishing work and are usually concealed beneath the surface. A *double-head nail* has two heads, one above the other. These are used for scaffolding or forms—work that must be disassembled later. The nail is driven only to the lower head; the upper head remains above the surface so that it can be grasped by a claw hammer for easy removal.

These are the basic woodworking nails, but there are many other special types and variations. *Masonry nails* may be either cut steel, as mentioned above, or specially hardened wire. Rust-proof *aluminum nails* are used for exterior applications, such as on certain types of siding; *galvanized steel nails* are used for the same purpose. *Roofing nails* have large heads to hold soft asphalt shin-

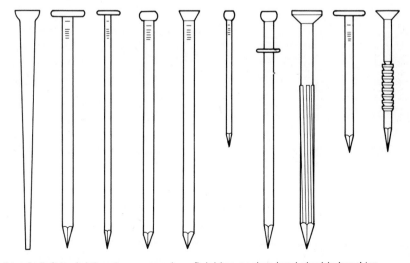

Nail types (left to right): cut, common, box, finishing, casing, brad, double-head (or duplex-head), masonry, roofing, drywall.

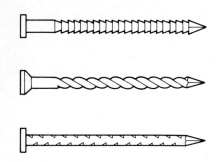

Ringed shank (top), spiral shank (center), barbed shank (bottom).

Nail Size and Length

Penny size	Nail length
2d	1 inch
3d	1¼ inches
4d	1½ inches
5d	1¾ inches
6d	2 inches
7d	2¼ inches
8d	2½ inches
9d	2¾ inches
10d	3 inches
12d	3¼ inches
16d	3½ inches
20d	4 inches
30d	4½ inches
40d	5 inches
50d	5½ inches
60d	6 inches

Approximate Number per Pound of Various Nails*

Nail size	Common nails	Box nails	Casing nails	Finishing nails
2d	890	1,010	1,010	1,380
3d	590	640	640	895
4d	318	440	440	605
5d	275	410	410	530
6d	190	240	240	322
7d	165	210	210	270
8d	106	145	145	200
9d	95	135	135	176
10d	72	95	95	130
12d	64	88	88	118
16d	48	72	72	92
20d	32	52	52	
30d	24	46	46	
40d	18	35		
50d	15			
60d	11			

* Based on average counts. Actual numbers may vary.

gles securely. *Dry wall nails*, with slightly smaller heads, are used to install wallboard.

In addition to variations in head shapes and sizes, nail points also vary. In general, the sharper the point, the greater the holding power of the nail. However, a sharp point is more likely to split the wood than a dull one, because the sharp point cuts its way through the wood fibers, rather than pushing them aside. Some types of nails are designed with blunt points to forestall splitting.

The shanks of some nails are ringed, spiraled, or barbed to increase holding power. Others may be coated with resin or cement. These nails are not recommended for use where they may have to be removed —they are very tenacious!

The "penny system" for sizing nails originated in England. The letter "d" was the designation for the English penny. In ancient times, the same abbreviation was used to indicate a pound in weight. Nails were weighed by the thousand, so if 1,000 nails totaled 12 pounds, they were 12d, or 12-penny, nails. Smaller nails might be 8d, 4d, and so on. The penny system has endured, although today it refers only to nail length. For example, a 2d nail is 1 inch long; the length increases ¼ inch for each higher number up to 10d; then the penny system gets a little more complicated (*see Nail Size and Length* chart).

Spikes are larger, thicker versions of common nails, overlapping some penny sizes—generally from 30d to 60d. Spikes longer than 6 inches are described by their actual length, ranging up to 12 inches.

Most nails are sold by the pound and, of course, the larger the nail, the fewer nails per pound. The type of nail also makes a difference in weight.

Nailing

The major problem in nailing is the danger of splitting the wood with the nail. Hardwoods are especially susceptible to splitting, as are such softwoods as white cedar, Douglas fir, and eastern hemlock. When there is a risk of splitting—for example, when nailing near the edge or end of a board—you can reduce the danger by blunting the tip of the nail with a file or by striking it with a hammer before driving. Rubbing the tip and the shank of the nail with soap or wax also helps. Or you can drill a pilot hole slightly smaller in diameter than the shank of the nail.

When nailing together two boards of different thicknesses, nail through the thinner one into the thicker one whenever possible. Ideally, the nail should extend into the thicker piece a distance equal to three times the thickness of the thinner piece—but this is not always feasible, since the tip of the nail might extend all the way through the thicker piece. For example, if you are nailing a 1 × 4 (¾ inch thick) to a 2 × 4 (1½ inches thick), a 7d nail (2¼ inches long) would seem to be just right. But if the nail is driven all the way in, chances are the tip would exit the other side. If this is acceptable for your project— fine. If not, use a 6d nail.

Face nailing is used when joining two pieces face-to-face. In *end nail-*

Toe nailing makes a strong butt joint.

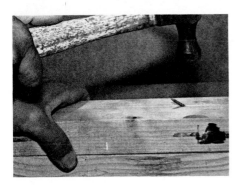

Clinching—bending over nail end.

Blind nailing is an effective technique for concealing a nail completely. Start by carefully prying up a sliver of wood: don't break it off.

Carefully drive a finishing nail beneath the sliver and countersink the head with a nail set. Be careful not to hit the sliver.

Use carpenter's glue to reattach the sliver in place. Press down the sliver and wipe away any excess glue. The nail is now hidden.

ing the nail is driven through the face of one piece into the end of the other. Both face and end nailing are used in accessible areas. *Toe nailing*— driving the nail in at an angle—is used where other methods cannot be. When nails are driven into the same place at opposing angles a strong joint is formed. Even stronger is *clinching*, driving a long nail through both boards, bending over the protruding end, and hammering it into the surface. Clinch nails when strength is important and looks are of no consequence.

On finished surfaces nails are driven below the surface of the wood with a nailset; the resulting holes should then be filled with wood putty or some similar substance to conceal the nail heads. An even more effective way to hide nail heads is *blind nailing* shown in the sequence of pictures at right.

For finish work that does not require the moderate strength of a finishing nail, use brads but don't drive them with a carpenter's hammer. The lighter tack hammer will avoid bending the brads.

Corrugated fasteners

Sometimes called wiggle nails, *corrugated steel fasteners* are often used to reinforce glued joints, such as butts and miters. Available in 1/4-, 1/2-, and 1-inch sizes, these fasteners have one sharpened edge—straight for hardwoods and sawtooth for softwoods. The fasteners are driven in diagonal to the grain to avoid splitting. Hammer blows should be distributed evenly across the top edge of the fastener until it is flush with the surface of the wood.

Screws

Wood screws are seldom used in structural carpentry. Framing is so designed that screws have little—if any—advantage over the cheaper, quicker, and easier-to-use nails. In woodworking, however, screws find many uses. They have better holding power than nails, and they can be tightened to draw the pieces being joined firmly together. Properly driven, they have a neat appearance. And they can be withdrawn without damaging the wood surface—this is important in such items as stereo cabinets, where access panels must be provided for cleaning and repairs.

Common wood screws are made of steel or brass. The screw shank is threaded from the point up, approximately two thirds its length; the upper third is smooth to the head. *Flat head screws* are used where the screw must be flush with the surface of the work. *Round head screws* protrude above the surface, making them easier to withdraw; they are also somewhat decorative. *Oval head screws* combine features of flatheads and roundheads; they are partly countersunk in the work, and they protrude slightly above the surface. Most screw heads have either a single slot across the entire width or a recessed, X-shaped slot (Phillips head).

Wood screw length is measured from the point to the widest part of the head. On a flat head screw, this is the overall length; on a round head, it is to the bottom of the head. Lengths

Corrugated fasteners are especially useful for reinforcing corners. One edge is sharpened and the fasteners are driven in like nails.

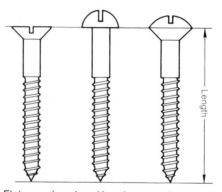

Flat, round, and oval head screws: length measured to wide part of head.

Slot-head screw and screwdriver (left), Phillips head and screwdriver (right).

range from 1/4 inch to 6 inches and are graduated by eighths of an inch to 1 inch, by quarters of an inch from 1 to 3 inches, then in 1/2-inch increments to 6 inches.

Screws also vary in body diameter, expressed as a *gauge number* from 0 (about 1/16 inch) to 24 (about 3/8 inch). Depending on length, screws are usually available in a variety of diameters. The lower-numbered (thinner) screws are used for fastening thin wood or where there is a danger of splitting, as when driving a screw into the edge of a board. The higher-numbered screws are used where greater strength is required.

Using screws

Whenever possible, select a screw long enough so that two thirds of the shank (the entire threaded part) will penetrate the lower board. For driving small screws in some softwoods, a starter hole can be made with an awl. For hardwoods and for larger screws, make pilot holes with a drill. Never attempt to start a screw by striking it with a hammer. This will damage both the work and the screw head.

Countersink and counterbore bits for electric drills allow the drilling of pilot holes in a single operation. The holes can also be made with standard twist drills, in either a hand or an

Screw Sizes: Lengths and Gauges

Screws are designated by both length and diameter. Length is designated in inches. Diameter is designated by a gauge number. Lengths available run from 1/4 inch to 6 inches. Gauges available are 0 (1/16 inch) to 24 (3/8 inch). The label on the box of screws might read 1x6, meaning the box contains 1-inch screws of No. 6 gauge. Most stores carry all standard lengths of screws in appropriate gauges. Most common gauges are Nos. 2 through 16. The heavier the work required of the screw, the larger the gauge should be.

Length in inches	Gauges Available																	
	0	1	2	3	4	5	6	7	8	9	10	11	12	14	16	18	20	24
1/4	X	X	X	X														
3/8		X	X	X	X	X	X	X										
1/2			X	X	X	X	X	X	X									
5/8			X	X	X	X	X	X	X	X	X							
3/4				X	X	X	X	X	X	X	X	X						
7/8					X	X	X	X	X	X	X	X						
1						X	X	X	X	X	X	X	X					
1 1/4						X	X	X	X	X	X	X	X	X				
1 1/2						X	X	X	X	X	X	X	X	X	X			
1 3/4							X	X	X	X	X	X	X	X	X	X		
2							X	X	X	X	X	X	X	X	X	X		
2 1/4								X	X	X	X	X	X	X	X	X		
2 1/2										X	X	X	X	X	X	X		
2 3/4											X	X	X	X	X	X		
3											X	X	X	X	X	X		
3 1/2												X	X	X	X	X	X	
4													X	X	X	X	X	
4 1/2														X	X	X	X	
5														X	X	X	X	
5 1/2															X	X	X	X
6															X	X	X	X

Screw Pilot Hole Sizes

Screw size (gauge)	Body hole (shank) in inches	Lead hole (threads) in inches
0	1/16	
1	1/16	
2	3/32	1/16
3	7/64	1/16
4	1/8	5/64
5	1/8	5/64
6	9/64	3/32
7	5/32	1/8
8	11/64	1/8
9	3/16	5/32
10	3/16	5/32
11	7/32	5/32
12	7/32	5/32
14	1/4	3/16
16	17/64	7/32
18	19/64	1/4
20	11/32	19/64
24	3/8	21/64

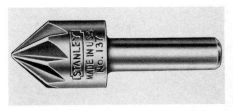

Countersink bit for an electric drill. This is used after the pilot hole is drilled to get head flush with surface.

electric drill. Use a bit with the same diameter as the shank body to drill a hole through the first piece or to a depth one third the length of the screw. Then drill a hole in the second piece slightly *smaller* than the diameter of the threaded part of the screw. The depth of this hole should be half the threaded length of the screw in softwoods, almost the full length in hardwoods. When flat head or oval head screws are used, countersink the hole at the surface. Countersink bits are available for both hand braces and electric drills. When the screw is to be driven beneath the wood surface so that the hole can be plugged to conceal the screw head, counterbore a hole slightly larger than the head.

Always start a screw perpendicular to the surface, and keep it straight while driving. If the pilot holes are straight, this should not be a problem. When the screw is difficult to turn, as is often the case in dense hardwoods, lubricate the threads by rubbing them with wax to ease the way. Always use a screwdriver that fits the width of the screw slot exactly. Hold the screwdriver in perfect alignment with the screw shank so that it doesn't slip out of the slot.

Occasionally, usually after a screw has been withdrawn and re-driven into the same hole a few times, the hole will become enlarged so that the screw threads no longer grip. Try stuffing the hole with a few wooden toothpicks or matchsticks (without the heads) dipped in white glue. This will usually hold the screw. If it doesn't, bore out the hole large enough to hold a dowel. Cut a dowel

to proper length, chamfer one end, and dip it into white glue. Drive the dowel into the hole with a mallet, chamfered end first. Let glue dry, then drive the screw into the dowel.

How to counterbore a screw

Counterboring involves drilling a hole deep enough to recess the head of the screw sufficiently so that it can be covered by wood putty or a plug.

STEP 1
Pre-drilling the screw holes
Drill a hole equal to the length of the screw plus the depth of the counterbore. If you plan to fill the hole with wood putty, a 1/4-inch counterbore is plenty; if you will be installing a plug, then make the counterbore depth 1/2 inch. The hole diameter should equal

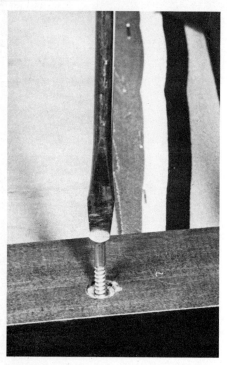

When driving or removing screws, use blade same width as slot.

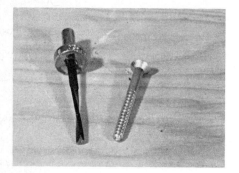

The Stanley Tools *Screwmate* countersinks and counterbores at once.

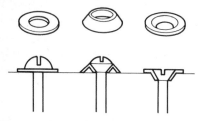

Screw washers (left to right) for round, oval, and flat head screws.

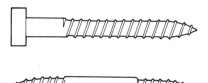

Lag screw (top), dowel screw (bottom); dowel screw is completely internal.

that of the shank without the threads.

STEP 2
Drilling the counterbore
To finish, use a bit the same diameter as the screw head to drill the counterbore.

If you expect to be making a fair number of counterbores of a particular size, you should invest in a screwmate. This relatively inexpensive bit drills the pilot hole, shank hole, and counterbore in one operation. Screwmates come in various common screw sizes. For example, if you will be counterboring No. $10 \times 1\frac{1}{4}$-inch screws, then you would buy a screwmate of the same size.

Screw washers
Screw washers are sometimes used to provide a greater bearing surface and to prevent marring the wood surface if a screw must be withdrawn frequently. The washers should be matched to the screw number. *Flush washers* are used for flat head screws; *flat washers*, for round head screws; *countersunk washers*, for oval head screws.

Lag screws
Lag screws, also called lag bolts, are thicker than ordinary wood screws and are used for heavy-duty fastening. They range in length from 1 to 12 inches, and in diameter from $\frac{1}{4}$ inch to 1 inch. Lag screws have a square or hexagonal head with no slot; they are driven by turning with a wrench.

Dowel screws
Dowel screws are threaded on both ends and are used in much the same way as their namesakes in some joints. Carefully aligned starter holes are drilled in the two pieces to be joined. The dowel screw is threaded into one of the holes. The mating piece is then twisted onto the protruding end of the screw.

Bolts
Used in heavy construction, as in joining timbers to form a roof truss or securing a fishplate reinforcement to a butt splice between two pieces of lumber, *bolts* are stronger fasteners than either nails or screws. They may also be used for such purposes as attaching table legs.

Bolts come in a wide array of types and sizes. Most hardware stores stock them in diameters of $\frac{3}{16}$ to $\frac{3}{4}$ inch, and in lengths from $\frac{3}{4}$ inch to 18 or 20 inches. As with screws, bolt length is measured from the tip of the shank to the widest part of the head. Bolts with square or hexagonal heads are measured to the bottom of the head. Some bolts are threaded along the entire shank, others only part way.

A *carriage bolt* has a round, unslotted head with a square shoulder just below it. When the bolt is drawn into the wood, the shoulder prevents it from turning as the nut is tightened on with a wrench. A *machine bolt* has a square or hexagonal head that is held with one wrench while the nut is tightened on with another wrench. *Stove bolts* come in small sizes only. They have flat, round, or oval slotted heads. A *truss head*—a flattened-out round head—is also available. *Hanger bolts* have wood screw threads on one end and machine threads to receive a nut on the other. They are designed for use where material that is attached to wood must be unfastened occasionally. The wood-screw end is driven into the work, with the machine-thread end protruding. The joining piece is fitted over the protruding bolt, and a nut is tightened on with a wrench.

Bolt length should be such that the end of the bolt extends slightly beyond the face of the tightened nut. For example, to join two pieces of 2-inch lumber ($1\frac{1}{2}$ inches each), a $3\frac{1}{2}$-inch bolt would be used.

Square or hexagonal nuts are considered a part of the bolt and are usually supplied with them. For quick fastening and unfastening, wing nuts may be purchased separately. For decorative purposes, or where the exposed end of the bolt might present a hazard, a cap nut may be used to cover the bolt end.

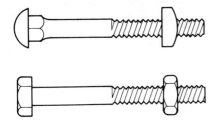

Carriage bolt (top), and hex head.

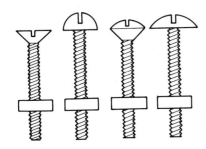

Stove bolts (left to right): flat, round, oval, and truss head.

Hanger bolt has machine screw threads on one end, wood screw on the other.

Nuts (left to right): square, hexagonal, wing, and cap.

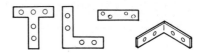

Metal plates (left to right): T, corner iron, mending, corner brace.

Where vibration may loosen the nut, use a lock washer with the flat washer.

Metal plates

Metal plates of various types are used to reinforce joints when appearance is not a consideration. *Flat, straight mending plates* range in size from $5/8 \times 2$ inches to $1 1/4 \times 8$ inches. *Flat corner irons* run from $3/8 \times 1 1/2$ inches to 1×6 inches. *Inside corner braces* come in sizes from $1/2 \times 1$ inch to $1 1/8 \times 8$ inches. *T-plates* are available from $1/2 \times 2 1/2$ inches to 1×5 inches, with the stem and arms of the T the same length and width.

Plates are made of either steel or brass, with holes countersunk for installation with flat head screws. If they are not provided with the plates, screws should be sized so that they fit flush with the surface.

Flat washer (left), and spring lock washer (right).

Holes for bolts may be drilled slightly larger than the diameter of the threads; for a snug fit, drill a hole the same size. Do not drill holes undersize and attempt to drive bolts through with a hammer. This will certainly clog the threads, and it may damage the bolt and the work.

Washers

Flat washers are often used between the bolt head and the wood, between the nut and the wood, or both. The washers spread the bearing surface and help protect the wood surface.

Hollow-wall fasteners

Toggle bolts may be used to fasten materials to walls between studs, to hollow-core doors, and to unfilled concrete blocks. These bolts consist of a threaded shank and a threaded, spring-winged toggle, and they come in various sizes. Inside the wall cavity, the wings will spring open—spreading the load over its opened length.

To install, drill a hole into the wall —you need a diameter large enough to admit the toggle with its wings folded against the bolt. Then drill a hole the same diameter as the bolt shank through the material to be attached to the wall. Remove the toggle from the bolt and pass the bolt through this hole. Then thread on the toggle and fold the wings.

Toggle bolts cannot be withdrawn from a wall without losing the toggle inside. *Expansion anchors* (familiarly known as Molly bolts, a tradename) can be removed and replaced,

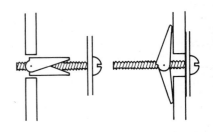

Toggle bolt wings are folded for insertion through a drilled hole (left), then open once in place (right).

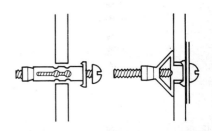

Expansion anchor is inserted in drilled hole (left); tightening pulls anchor body tight against wall.

Collar allows bolt to be withdrawn without losing expansion anchor.

Woodworking Adhesives

Adhesive	Preparation	Application	Setting Time	Moisture Resistance	Color (when dry)	Typical uses
Acrylic	two parts (liquid and powder); mix amount needed just before use	paddle, brush	15–20 minutes	waterproof	light tan	outdoor furniture, boat building
Aliphatic Resin	ready to use	squeeze bottle, brush	8–10 hours	low	clear	furniture building, cabinetry
Casein	powder; mix with water	brush	6–8 hours	high	milky white	general woodworking; good on oily woods
Contact cement	ready to use	brush, roller	bonds on contact	very high	clear	bonding veneers, laminates
Epoxy	two liquid parts; mix amount needed just before use	paddle, brush, roller	read label directions	waterproof	usually clear	not for general woodworking, but excellent for bonding dissimilar materials, e.g., metal to wood
Hide Glue (flake)	soak in water; heat	brush	8–10 hours	low	light brown	general furniture work
Hide Glue (liquid)	ready to use	brush, paddle	8–10 hours	low	honey	general furniture work
Hot-Melt	ready to use	electric glue gun	5–15 minutes	depends on type of glue	depends on type of glue	quick repairs
Mastic	ready to use	caulking gun	read label directions	low	dark brown	bonding paneling to furring strips
Polyvinyl Acetate (PVA, white glue)	ready to use	squeeze bottle	8 hours	low	clear	general woodworking
Polyvinyl chloride (PVC)	ready to use	squeeze tube	read label directions	high	clear	furniture repairs, crafts
Resin	powder; mix with water	brush, paddle	16 hours	high	clear	wood projects in high-moisture areas; not good for oily woods
Resorcinol	two parts (liquid and powder); mix amount needed just before use	brush, roller, paddle	5–15 hours, depending on temperature	waterproof	dark red	outdoor furniture, boat building
Urea-formaldehyde	powder; mix with water	brush, paddle	10–14 hours	high	light brown	furniture; cabinetwork

making them the choice for hanging items that may have to be taken down on occasion. The anchors are inserted through drilled holes in a wall. As the bolt is turned, prongs are forced outward to form a broad shield to hold the bolt and provide a wide load-bearing surface. The bolt is then withdrawn, inserted through the item to be attached, and redriven into the wall anchor.

The hollow-wall anchor selected for any job should have a shank as thick as the wall material. This may range up to 1¾ inches thick, but in most homes an anchor with a ⅜-inch or ½-inch shank will be sufficient, because these are the most common wall thicknesses.

Adhesives

The finest joinery is accomplished with adhesives, either used alone or in conjunction with screws, nails, or corrugated fasteners. Practically all cabinetwork employs glued joints; if properly done, such joints can be stronger than the wood itself.

The quality of a glued joint depends on several factors:

- kind of wood (generally, hardwoods are more difficult to glue than softwoods; heartwoods are more difficult than sapwoods)
- moisture content of the wood (kiln-dried wood is generally preferred)
- type of joint
- the match between joined surfaces (it should be as precise as possible)
- type of glue and method of preparation and application
- degree and duration of pressure applied while the adhesive sets

There are many woodworking adhesives available, and new ones, or improvements or variations of existing ones, are constantly coming into the marketplace. But there is no one glue for all purposes.

Selecting the right glue is half the job. Applying it properly is the other half. But neither the right glue nor the proper application means much unless you clamp the work and give the glue all the time it needs to dry. A well-made glue joint is strong, neat,

and durable. A poorly made joint is sloppy looking and comes apart quickly under stress.

It may sound like gratuitous advice, but always read the label on any new adhesive you buy. Most of us have been using glues so long that we think we know how they all work, so we rush ahead to use a new glue as we used the older ones. But newer glues sometimes require different application techniques. Read the label to find out. Manufacturers want you to get good results with their product, and the instructions are designed to help you get those results.

Gluing techniques

In gluing, more is *not* better. Some people feel that if a little glue holds firmly, then a lot of glue holds even better. In fact, the opposite is true. Too much glue makes a weak joint, chiefly because glue in itself is not a strong substance—not nearly as strong as the wood it bonds.

If you placed two pieces of wood ⅛ inch apart, filled the space between them with glue, and let it set, you would have a weak joint that could be easily broken. On the other hand, if you applied a thin coating of glue to each of the surfaces, then clamped them firmly as they dried, you would have a joint that in many cases was stronger than the wood itself. If you tried to break the joint apart, the wood on either side of the joint would probably fracture before the joint broke.

And therein lies the secret of making a good glued joint. Apply thin coats of glue, clamp securely, and allow ample drying time.

Glues do not adhere well to very smooth surfaces but work best when they can grip something. For this reason, you should roughen any smooth surface slightly before applying glue.

Let's look at an example: a chair rung has pulled out, and you want to glue it back in place. The surface of the rung is probably very smooth. You can make a better glue joint if you roughen the surface of the rung where the glue is to be applied. Try a

Before gluing a joint, test its fit. For lasting strength, the pieces should be in contact at all points along the joint.

To control the amount of glue and prevent drips, apply glue with a brush or stick. Plywood end grain absorbs glue: apply one coat, wait, then apply a second.

Tightly clamp glued joints after checking for fit and squareness. In cases when nails or screws are also used in the joint, clamping is unnecessary.

few passes with coarse sandpaper or scrape a little with the blade of your pocketknife. Then brush on a thin coat of glue and reinsert the rung into its hole.

Clamping a glue joint is essential, but a careless use of clamps can create problems. Always use a pad of some kind between the clamping surfaces and the face of the wood, so that you do not dent or mar the work. Thin little shims of wood make good pads. This advice is especially applicable when using C clamps.

The final joint should be neat and clean. So after setting the clamps in place, wipe away any excess glue with a damp cloth. Watch for any later drips or runs; wipe these off, too. There is absolutely no holding value in glue on the outside of a joint. The glue does all of its work on the two butted surfaces within the joint.

Glue types

Here are some of the properties of the more common adhesives you will encounter in carpentry and woodworking. These glues should take care of most of your needs, so stock up on them. Consult the *Woodworking Adhesives* chart for more information on these and other glues.

Polyvinyl glues. The white creamy glues that come in plastic squeeze bottles (Elmer's is a leading brand name) are polyvinyls. They are inexpensive, set in an hour or so, and work in just about every furniture, craft, or woodworking project. Polyvinyls dry clear and won't stain any wood. They do have one drawback— water will soften them after they have set. Don't use a polyvinyl to glue the sides of a fishtank or to seal the edge of a bathtub.

Resorcinol and formaldehyde glues. You mix these just before using. The resorcinols come in two parts, a resin and a powder; the formaldehydes come as powders that you mix with water. Both are good for from two to four hours after mixing, and both make very durable joints. The resorcinols are waterproof, but the formaldehydes are not. Follow the manufacturer's instructions on drying. Usually the time ranges from three to twelve hours under clamps. Use both types at temperatures over 70° F. A small problem: these glues are brown in color and will stain light-colored woods.

Contact cement. This is a stronger version of the familiar rubber cement. Contact cement is used mainly to apply veneers and to bond plastic laminates to wood for table and counter tops. The correct way to use a contact cement is to apply a thin coating to both surfaces and allow both to dry. Then press the surfaces together. Be careful how you do this, since the two surfaces will stick together instantly on contact, and you won't be able to pull them apart with a tractor. Since they can't

be adjusted after contact, be sure to align them before you put them together. Practice aligning techniques on old wood before trying it on a project.

Chemical-based contact cements have a strong odor and must be used in well-ventilated areas. The water-based types are more expensive but are safer to use.

Epoxies. No adhesive is tougher than an epoxy. Epoxies come in two parts, a resin and a hardener, which must be mixed just before you use the glue. By all means, read the label before mixing, because it will tell you the correct proportions of resin and hardener to use. If you mix the wrong proportions, you may end up with a sticky, nondrying mess.

Once it has set, epoxy cement resists almost everything, from water

to gasoline to solvents. An epoxy is the only adhesive as strong or stronger than the material it bonds. It is the one adhesive you *can* put between those two pieces of wood 1/8 inch apart and expect a strong joint. For this reason, epoxies are sometimes used to fill large cavities. The dried epoxy can be machined, sanded, and shaped, if necessary. For this reason, too, an epoxy is not applied in a thin coat as other glues are, and it should not be tightly clamped while setting. Clamping may squeeze out too much.

Epoxies must be used in warm temperatures, since the warmer the air around them, the faster they set. Setting time varies considerably from brand to brand so, once again, read the directions before using.

7 MOLDING AND TRIM

Whether for finishing a room, installing paneling, coordinating new cabinets or bookcases, or just adding a nice decorative touch to your work, you will want to investigate the many kinds of molding available. Many standard shapes, discussed below, have evolved over the years for various uses, and most can be adapted to your special needs. Also, you can have custom molding made for you in a cabinet shop, and you can create your own with a router. Even flat pieces of wood can give decorative highlights to a piece of work, as shown in this kitchen cabinet installation.

Molding basics

In essence, molding gives a project a decorative, cut-from-solid look, or it softens corners such as the junction of floor and walls or of ceiling and walls. All that is involved in achieving these effects is to attach molding to the face of the door or drawer or cabinet or wall with brads and glue. If your joints are tight and your nails well hidden, only extremely close scrutiny will reveal that molding had been used in the three-dimensional design.

Moldings come in many different configurations. The molding may be made of softwood or hardwood: cedar, pine, fir, larch, and hemlock are commonly available at lumberyards. As with lumber and plywood, however, availability depends on the particular region you live in.

You can obtain standard, unfinished molding in any length up to 16 feet, with the lengths increasing in 2-foot increments. You may also obtain it in odd-length increments—3-foot, 5-foot, and so on—if this suits your purposes better. Also, buying random lengths is less expensive.

Molding is available in nominal and actual sizes, similar to board sizing, with the actual size being smaller than the nominal. For example, a piece of case molding nominally 3 inches wide would be approximately $2\frac{5}{8}$ inches wide in actual size. And, as with board, this is something you must keep in mind when planning your projects' dimensions.

Techniques for milling molding are not perfect. Molding of the same nominal sizes will have fractional dif-

The use of moldings on walls and as decorative picture frames greatly enhances the character of this small room. The possible effects are endless.

Casing, in any of various configurations, is used to trim inside and outside door and window openings.

Base molding is installed where floor and walls meet, protecting walls from bumps, and forming a visual foundation.

Lattice (left), screen (center), and astragal moldings have decorative as well as useful functions.

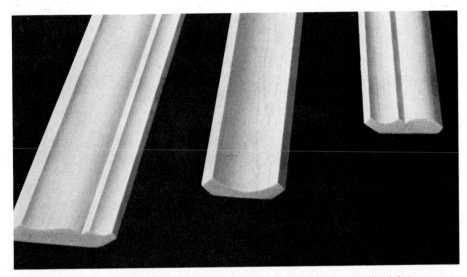

Crown (left), cove (center), and bed moldings (right) are all used at wall-ceiling joints.

Half-round (left), quarter-round (center), and round moldings (right). Round is used for closet poles.

ferences in the measurements. It is therefore suggested that you buy all your molding from the same mill lot and that you hold the pieces up, superimposing one upon the others, to make sure that they are all the same size. Fractional differences in sizes can detract from the even appearance and professional-looking effect that you desire.

Standard molding

A number of standard molding styles have retained their popularity throughout the years. Following is a description of these basic styles. While most of these have been designed for specific uses—such as base molding for walls—there is no prohibition against using them on a bookcase or cabinet or wherever you like if the molding fits into the overall design plan.

Base shoe. This is normally used with base molding where walls meet floors.

Base molding. The molding used along the bottom of walls, base molding covers the gap between base shoe and floor. It is available in many styles.

Quarter round. In cross-section, this looks like a quarter of a circle. It is most commonly used to finish corners.

Half round. In cross-section and in the smaller sizes, this looks like a half a circle. In the larger sizes half round molding has an oval shape.

Casing. This is probably the molding most frequently used around the house. It is used to trim edges of doors and windows, and is a very popular molding for finishing cabinets.

Mullion. This fluted molding is used as vertical trim between windows.

Chair rail. This molding was originally designed to protect walls from the backs of chairs. Today it is used for decorative purposes, installed horizontally midway up the wall.

Panel molding. This type of molding can come in handy if you cover anything—from a wall to cabinetry —with paneling. It will hide the seams and will make the panel edges appear inset.

Corner guard. This is a decorative molding, available in a V shape, for covering outside corners of pieces. An inside corner guard is also available.

Astragal. Another decorative molding, it is normally used to divide windows.

Cove. This is the standard molding used to cover ceiling joints; it is sometimes added to cabinets.

Specialty moldings

In addition to standard moldings, there is a wide variety of special-use moldings available; they come with cuts made in the surface. Visit a few lumberyards and see what is in stock so you can determine if the available moldings will fit your project. Like most wood moldings, specialty molding will be purchased unfinished.

Prefinished moldings

Some moldings come prefinished. There are plastic moldings, which come in white, as well as wood-toned ones that have their faces covered with a vinyl that is imprinted with various wood patterns. There are also moldings designed to be used with paneling; they mask the top and bottom edges and spaces between panels.

Working with molding

The miter is the most common joint required for joining pieces of molding, so a good back saw and a miter box are essential. Once the molding has been cut, you can attach it with brads and glue. Use a minimum amount of glue: just enough so that it will adhere to the material but not so much that it will squeeze out when the piece is pressed in place. Above all, you want to keep the glue off the molding face. Removing the glue can affect the wood in terms of the way it colors the finish you are using.

Here is a sampling of the decorative effects that can be achieved on the panels of doors and cabinets with the use of different moldings.

Use as few brads as possible to hold the molding in place. Use a brad driver to push the brads in place, then a nailset to sink the heads of the nails slightly below the surface; then fill with wood filler and sand smooth.

Sometimes the molding must be sanded to remove minor imperfections. Do this very carefully so that you will not change the shape of the material.

Molding, not so incidentally, can make up for some imperfections; if you have left gaps in joints the molding can be used to cover them neatly.

The router and molding

The versatility of the router is due to its variety of bits. The accompanying illustration gives an indication of this. In brief, however, the bits come in a range of diameters and shapes to help you make just about any cut you need. When the decorative cuts

Straight bits

Two flutes

Single flute

Grooving bits

Veining

Core box

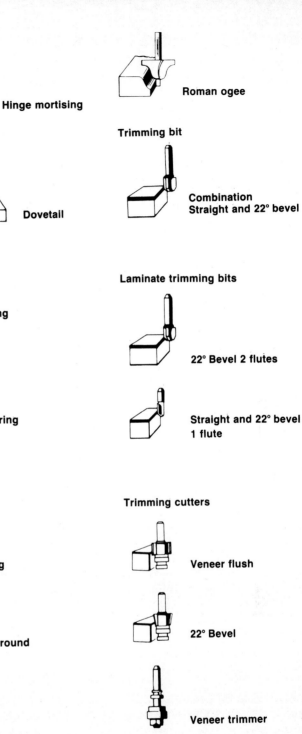

"V" grooving

Grooving bits

Hinge mortising

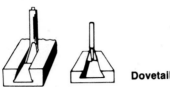

Dovetail

Rabbeting bit

Rabbeting

Decorating bits

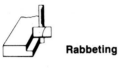

Chamfering

Cove

Beading

Corner round

Ogee

Roman ogee

Trimming bit

Combination
Straight and 22° bevel

Laminate trimming bits

22° Bevel 2 flutes

Straight and 22° bevel
1 flute

Trimming cutters

Veneer flush

22° Bevel

Veneer trimmer

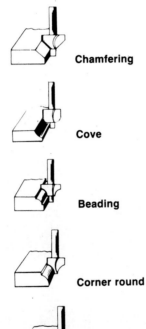

and grooves are combined with molding, then the possibilities for trimming a cabinet, bookcase, or other project increase dramatically.

The easiest way to make cuts or grooves in the face of a cabinet door is to use a jig of your own making with commercially available guides. The guides are simply rings of metal. They come in various sizes to accommodate various bit sizes, and they screw to the baseplate of the router with the bit projecting through them. The bits, of course, can be raised or lowered to cut to whatever depth you wish.

To use one of the guides, clamp your jig or board—a straightedge or a board cut to whatever shape you need—to the face of the material to be cut. Then run the router along the material with the ring riding against the jig or the board.

If you have to make a fancy cutout on the router, it is a good idea to

The look of molding can be achieved with a router directly on the edge of a piece of work. The bits shown here are only a sampling.

make a paper pattern first. Divide your paper into 1-inch squares—or whatever size is most convenient—then draw the outline you need on the squares. You can then cut out the

pattern and tape it to the wood as a guide, or tape the pattern in place, draw the outline from it, and then remove it.

8 FRAMING FUNDAMENTALS

First, a disclaimer. This chapter does *not* contain all you need to know about how to build a house or an addition to a house. If that is your intention, you will need far more information (on such topics as concrete, masonry, roofing, mechanical work, the myriad details of carpentry for specific situations) than we can possibly include here. We suggest that you visit your bookseller or library to find a book that addresses these topics, or is devoted exclusively to home building.

The purpose of this chapter is to impart the basic principles of structural framing. Knowing how your house is built will enable you to undertake many projects: creating (or filling in) door or window openings, finishing off attic or basement rooms, erecting built-ins, adding a dormer for a second-floor room, and countless repair jobs. As you gain experience and confidence, you will probably be able to graduate to the big time—framing an extension to your house, maybe even framing an entire house. Others have done it; why not you? After all, framing follows a logical progression—every element depends on a previously installed element.

Many older homes were built with balloon frames or braced frames. Braced-frame construction is an English import, dating back to colonial times. In the original version, heavy corner timbers extended from the foundation to a heavy beam at the roof line. Intermediate posts sup-ported another timber, or girt, at the upper-floor levels, which in turn supported floor joists and posts at that level. Balloon-frame construction, a later development, has studs extending from the foundation to the roof line. Joists are supported by a ledger board recessed into the studs. Balloon framing was widely used until about forty years ago, but today balloon frames and braced frames are rarely seen in new construction. Platform framing is by far the most common method now used.

In *platform framing* (sometimes called Western framing), the first floor is built on the foundation walls. This floor then serves as a platform to support the first-story walls, which in turn support the second-story floor (in multi-story homes), which serves as a platform for second-story walls—and so on. Platform framing is preferred because of its simplicity and relative economy (it does not require the very long posts and studs demanded by other framing methods), and also because shrinkage due to settling is more uniform throughout, and therefore less noticeable.

The platform

For a structure with a basement or crawl space, a concrete footing is poured below the frost line, then a concrete or concrete block foundation wall is erected on it, with anchor bolts embedded in the top about every 4 feet, protruding at least 2¼ inches. (These and all other dimen-

This apparently impenetrable forest goes together in a logical manner.

In platform framing, the floor is built first, the walls rise on the platform.

sions given in this chapter should be checked against local design considerations and building codes.) In concrete-slab construction, the slab itself serves as the platform on which the walls are built; anchor bolts are embedded around the perimeter.

Floor joists normally run across

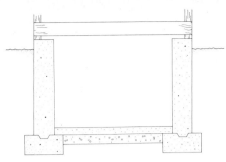

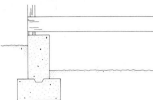

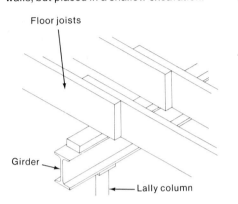

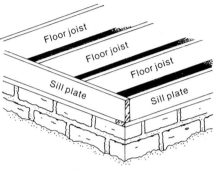

Full basement foundation with footings and masonry walls about seven feet high.

Crawl space foundation is constructed like a full basement with footings and masonry walls, but placed in a shallow excavation.

Slab-on-grade foundation with footings inserted into the ground and concrete slab laid directly on the ground.

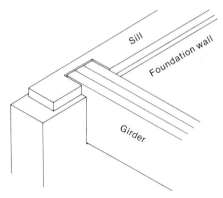

Foundation wall is notched so ends of girders sit flush with top of sill.

Girders may be steel I-beams supported by cement-filled lally columns.

The ends of the floor beams are anchored to box beams that sit on the sill.

the narrow width of the foundation. If the structure is no more than 14 or 15 feet wide (check codes), the joists may span the full width. For greater spans, a *girder* or beam is placed along the centerline to support the joists. The girder may be wood (usually three pieces of 2×8 or 2×10 nailed together) or a steel I-beam recessed into the foundation wall. A wood beam is set with its top $1\frac{1}{2}$ inches above the foundation, level with the sill plates. An I-beam is set flush with the top of the foundation, then a length of 2×4 or 2×6 is fastened on top of it, so that it is level with the sill. The girder is supported at 8- to 10-foot intervals by posts resting on concrete footings. The posts are either 4×6s or larger timbers or steel lally columns.

The *sill* or *sill plate* consists of lengths of lumber (usually 2×6s) laid flat on the foundation wall to provide a bearing surface for the floor joists. The sill is drilled to fit over the anchor bolts and laid in a bed of mortar or fiberglass sealer. Flat washers fit over the bolts, and nuts are tightened on to hold the sill securely. (Where there is danger of termite invasion, a metal shield is placed between the foundation and the sill.)

Floor joists are the main supporting members of the floor. They rest on the sill at the outside and on the girder at the inner end. Depending on the span and the floor load, joists are most commonly 2×8s or 2×10s (sometimes 2×12s) set on edge, usually spaced 16 inches center-to-center (o.c.), which means from the center of each board. Joists overlap at the girder, so this must be taken into account when laying out the positions on the sill plates. Where partitions on the floor above will run parallel to the joists, double joists are usually provided to carry the weight. If heating ducts or plumbing pipes will be run through the partition, the joists are usually spaced so that they will be on each side of the partition, rather than directly beneath it (unless this interferes with the 16-inch o.c. spacing pattern). Openings for basement stairways, crawl space access, chimneys, and the like require double joists and headers all around; but again, these should not alter the regular joist spacing. Header joists and end joists enclose the perimeter of the floor framing.

Cross bridging consists of diagonal pieces of wood (1×3s, 1×4s, or 2×2s) arranged in an X pattern between adjacent joists, midway between sill and girder. The purpose is to stiffen the floor structure, but recent research has questioned the benefits of cross bridging, indicating that it shows no significant ability to transfer loads after the subfloor and finish floor are installed. Still, cross bridging is required by many building codes. When used, it is nailed at the top first. The lower ends are left loose until the subfloor is laid; then they are nailed.

Solid bridging—usually a piece of lumber the same size as the joists— is often used between joists to provide a more rigid base for partitions above joist spaces. Solid bridging is also used where irregular spacing may make it difficult to install cross bridging.

Subflooring is nailed (sometimes glued) over the joists. Plywood is the material of choice for subflooring; it goes down quickly and covers large areas in minimal time. C-D Interior

Sometimes plywood subflooring is attached to floor joists with a special adhesive. Properly done, this makes a strong, rigid platform.

plywood is commonly used for subflooring. Panels of this grade carry a marking that gives allowable spacing of joists for various thicknesses of plywood. A marking of 32/16 means that spacing can be 32 inches if the panel is to be used for roof sheathing; 16 inches, for subflooring. The plywood should be laid with the better (C) face up. Grain direction of outer plies should be at right angles to the joists. Panels should be staggered so that end joints in adjacent panels fall over different joists (the ends must always be centered on a joist). Allow 1/8-inch spacing between panels at edge joints and 1/16-inch spacing at end joints over joists.

Boards (most commonly 1×6s or 1×8s with tongue-and-groove edges) are still sometimes used for subflooring. They may be laid either at right angles to the floor joists or diagonally across them. The latter method adds some rigidity to the structure by tying walls firmly together, but it also means that each board has to cover a greater span (approximately 20 inches) between joists than boards laid perpendicularly (which span only 14½ inches). This can be a source of floor squeaks later on. As with plywood subflooring, board ends must always fall over a joist, regardless of how the flooring is laid.

Framing a house

The following instructions are a condensed version of how to frame a house. They are written as if you were actually doing the job to give you a better idea of what is involved. Picture the steps as you read these instructions, and you will see just how easy framing is. When you are ready to frame a house yourself, you can use this outline as a quick refresher course. The chart on page 90 gives the sequence of nailing.

STEP 1
Framing the walls

After the floor platform is in place, the next step is wall framing. *Plates* (bottom and top horizontal members) and *studs* (vertical members) in conventional construction are usually of 2×4 lumber. 2×3s are sometimes used for non-load-bearing walls in attic and basement finishing.

On the subfloor, lay out and mark with a sharp pencil the locations of all exterior and interior walls and partitions. For one of the long exterior walls, select a 2×4 sole, or bottom plate (it may have to be more than one piece), and, using double-head nails, nail it temporarily to the subfloor with the outside edge of the 2×4 flush with the edge of the header joist. Mark the locations of all wall openings on the plate (use the roughing-in or overall dimensions for the particular doors and windows you will be installing). Mark the location of the first corner stud, then measure 15¼ inches from the corner and mark a line across the plate; this will indicate the edge of the first intermediate stud. Make an X ahead of

the line to indicate where the stud will be positioned. From this line, measure and mark the locations of the rest of the wall studs at 16-inch intervals (using the 16-inch tongue of the carpenter's square makes it easy). Mark locations of double studs at openings; do not alter the original 16-inch o.c. spacing, or you will end up with a lot of waste when you install sheathing and interior wallboard (both of which are in 4-foot modular panels). Extra studs at wall intersections should also be marked.

Select a lower top plate and temporarily nail it alongside the bottom plate. If more than one piece is needed because of the length of the plate, center the joint on one of the stud markings. Transfer all the markings from the bottom plate to the top plate, using the square laid across both plates to ensure accuracy.

You may be able to buy pre-cut studs at your lumberyard; but if they do not have the length you need, cut them all at one time. Stud length is determined by the height of the room, plus the thickness of floor and ceiling finishing materials, minus 4½ inches (for the bottom plate and the double top plate). For example, for a room that is 8 feet high, with vinyl floor tile and a gypsum board ceiling, you need: 8 feet + 3/8 inch (underlayment and tile) + 1/2 inch (ceiling) − 4½ inches. This gives you a stud length of 7 feet, 8⅜ inches.

Select a straight 2×4, measure it, and cut it to the exact length. Double-check the length, then mark it "stud pattern." Use this pattern to cut all other full-length studs for all walls and partitions.

Wall framing is most quickly and

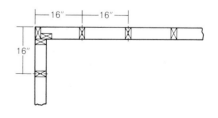

In wall framing, studs are spaced 16 inches center-to-center.

easily assembled flat on the platform, then tilted up into position. Remove the double-head nails holding the plates to the floor and set them on edge, one stud length apart, with the markings facing inward. Place the studs between the plates, then end nail the plates to the studs, making sure that the studs are aligned with, and on the X side of, the marks.

Headers, or *lintels*, are used to span window and door openings, making up the support otherwise provided by the studs. Headers are formed of two pieces of 2-inch dimension lumber set on edge, with spacers of ½-inch plywood scraps sandwiched between to make up the full wall thickness of 3½ inches. The following are the recommended lumber sizes for various spans:

Maximum span	Header (double)
3½ feet	2×6
5 feet	2×8
6½ feet	2×10
8 feet	2×12

Assemble the headers, then cut support studs to the height of the

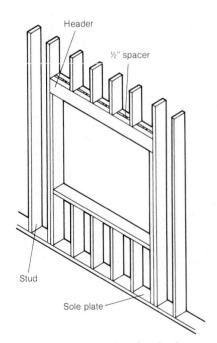

Framing for a window; door framing is similar without studs below opening.

Walls and partitions are usually assembled horizontally on the platform, then tipped up into place. This operation takes some muscle.

rough openings and nail them to the flanking studs. Place the headers in the openings on top of the supporting studs and nail them in place. Complete the framing with short studs between the header and the top plate where required, and with studs (these are called *cripple studs*) and sill below window openings.

Check the wall for squareness by measuring diagonally from corner to corner. When the wall is square, the two diagonals will be exactly the same length. With double-head nails, fasten lengths of 1×4 or 1×6 diagonally across the framing to hold it in square.

Tipping up the wall section is a job for at least two workers—three or four or more for very long walls. Begin with a long wall. Lift the top plate, taking care not to push the bottom plate off the edge of the platform. When the wall is upright, the workers hold it in place. Make sure that it is flush with the outer edge of the header joist and aligned at each end, then drive three or four nails along the length of the wall through the sole plate and subfloor into the

joists. Use a carpenter's level on several studs along the wall's length to make sure that the wall is perfectly plumb (vertical), then hold it in position by attaching temporary diagonal braces of 1×4 or 1×6 lumber to the studs and to 2×4 blocks nailed to the floor; use double-head nails here for easy withdrawal later. Complete nailing the sole plate to the subfloor and joists.

Assemble and erect the opposite long wall in the same fashion, then the two end walls and the interior walls and partitions. Make sure that all are plumb and nail them securely together at corners and intersections to form a rigid, sturdy framework. When all walls are in place, install the top plates. Overlap the lower top plates at corners and intersections to lock the structure securely together. Where, because of wall length, more than one piece must be used for a top plate, the joint should occur over a stud (but not over the same stud as a joint in the lower top plate). Leave the diagonal bracing in place until ceiling joists and rafters are installed.

The long walls are erected first, then the remaining walls and partitions are built and tied to the walls in place.

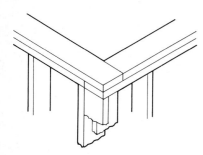

Top plates overlap lower top plates to securely lock corners together. This reinforces the connection made at the built up corner post below.

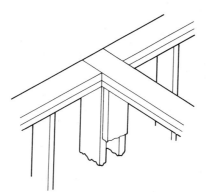

Where partitions intersect walls, the top plate extends over the wall.

STEP 2
Framing the ceilings

Once the exterior and interior walls are erected, plumbed, braced, and tied together by top plates, the ceiling framing is installed. In a full two-story house, the ceiling framing over the first story is essentially the same as the floor framing previously described; there is a main bearing wall in place of a girder. (In some layouts, such as the familiar L-shaped living-dining room, the bearing wall is not continuous, and a beam or girder carries the load over the opening, just as in first-floor construction.) Subflooring over the ceiling joists serves as the platform for second-story wall framing.

In one- and one-and-a-half-story structures, as well as on the upper story of multi-story homes, the ceiling framing serves the additional purpose of tying together opposite walls and roof rafters to resist the outward pressure imposed by the pitched roof. Layout of ceiling joists and roof rafters should be done at the same time, since joists and rafters must lap and be securely nailed together. Sixteen-inch o.c. spacing is common, although 24-inch and even wider spacing is sometimes allowed. Check local building codes in this regard; check also for required joist sizes. Where they must support a second floor, joists are usually 2×8 or 2×10 lumber. Where there is no

floor overhead, or simply a crawl space floor, 2×6 joists are often all that is needed.

Mark the locations of the ceiling joists on the top plates of the outer walls and the main bearing wall, and of the rafters on the outer walls. Since rafters on opposite sides of the roof should be aligned with one another, but the ceiling joists will overlap at the bearing wall, allow for a filler block the same dimension as the joists to be sandwiched between them at the bearing wall. Then place one ceiling joist on each side of the roof rafter, with the filler block occupying the space of the rafter at the center.

Cut ceiling joists to run from the outside edge of the outer walls and overlap at least 4 inches over the main bearing wall. (Joists at the end walls are butted, not overlapped.) Trim the upper corners on an angle at the outer ends where they must match the slope of the roof rafters. With a helper, set all the joists flat on the outer wall and bearing wall (this saves a lot of climbing up and down). Then, starting at one end, place them on end and toe nail to the outer wall. When all are nailed, move to the inner wall; nail the joists to the filler blocks; then toe nail them to the bearing wall.

Openings in ceiling framing are formed the same way as in floors. Where partitions run parallel with and between joists, nail 2×4 headers between the joists, then nail the headers to the partitions to hold them securely in place.

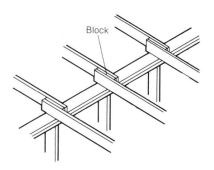

Blocks between joists allow rafters to align at ridge board overhead.

Cutting ceiling joists for roof pitch.

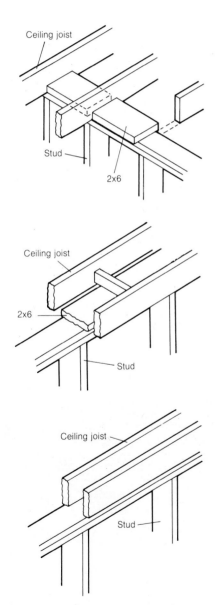

Nailing surface for ceiling is a necessity: (left) 2 × 6s between joists, (center) 2 × 6 nailer between joists, and (right) overhanging ceiling joists.

STEP 3
Framing the roof

Rafter size depends on several factors: the horizontal span (not the overall length of the rafter), the weight of roofing material (slate, for example, is considerably heavier than asphalt shingles), and the maximum likely wind and snow load. Again, check local building regulations.

The gable roof is the most common form of pitched roof and, except for a shed roof, the least difficult to build. Each pair of rafters is fastened at the top to a ridge board, commonly a 2 × 8 (for 2 × 6 rafters) or a 2 × 10 (for 2 × 8 rafters).

Rafter locations on the outer walls are usually laid out at the same time as the ceiling joists. If this was not done as described previously, mark the rafter locations on the top plates of the side walls. The first pair of rafters at each end of the structure will be flush with the outer edge of the end wall.

Mark the rafter locations on the ridge board. (If there is to be any gable overhang, allow for it in the length of the ridge board.) Since the outer rafters are end nailed to the ridge board, the ridge should be 1½ inches shorter than the overhang at each end. For all but the smallest structures, the ridge board will probably have to be spliced to achieve its overall length, but do not complete the splice yet. It is easier to erect it in shorter sections. Just make sure that the splice occurs between rafters, so that scabs can be nailed on both sides to strengthen it.

Except for the two end pairs, all rafters in a gable roof are cut to the same length and pattern. You must know the *slope*, or pitch, of the roof to lay out the pattern. This is expressed as the number of inches of vertical rise in 12 inches of horizontal run. For example, a 4 in 12 (or 4/12) roof rises 4 inches for every 12 inches of run; this is a common slope for one-story roof structures. A 10/12 roof would be much steeper, such as that on a one-and-a-half-story house.

Draw a full-size rafter pattern on your work platform (the subfloor),

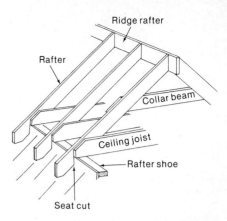

Rafters provide the main support for the roof. Usually they are notched so that one end ties into a rafter shoe; other end lies flat against ridge rafter.

showing the slope. From this drawing, determine the length of the rafters (including any overhang), the angle of cut at each end, and the location of the notch or *seat cut* where the rafter sits on the wall plate. Bear in mind that the rafter will butt against the ridge board at the top, so deduct a vertical ¾ inch from that end of the rafter to allow for the ridge. From this pattern, lay out one pair of rafters, marking top and bottom angles and seat cut locations. After cutting, check the fit by setting the rafters up at floor level. When you are satisfied with the fit, mark them "rafter pattern." Use this pattern to cut the remaining rafters. For the two sets of end rafters (*fascia rafters*), add the ¾-inch half-width of the ridge board and eliminate the seat cut (unless there is no gable overhang and these rafters, like the others, will sit on the wall).

It's best to have a few helpers, at least during the beginning stages of rafter erection. Once the ridge board and a few rafters are in place, two people can handle the job comfortably. Tack temporary 2 × 4 supports to the top of the main bearing wall or to ceiling joists to hold the ridge board. You should place some sheets of plywood or several boards across the ceiling joists for safer footing. Lift the ridge board sections onto the ceiling framing. Lean rafters up against the house or place them on the outer and bearing walls. With one person holding the ridge board at the approximate height, nail the

Installation of rafters is best done by at least three workers: one to support the ridge board, one to nail to ridge, one to nail to outer wall.

Sheathing is placed over the walls, sheets of plywood that continue to the gable itself.

the stud bottoms will be flush with the top wall plate. Nail the cripples to the plate and the end rafter.

Fascia rafters are supported by *outriggers*—boards set into notches in the end rafters and fastened securely to the second pair of rafters. Nail the fascia rafters to the outriggers and end nail them to the ridge board.

STEP 4
Sheathing the walls

Wall sheathing is applied over the wall framing to provide a base for the exterior finish material. The common types of sheathing are boards and panel materials (plywood, structural insulating board, gypsum board).

Board sheathing is normally of 1×6, 1×8, or 1×10 lumber, with tongue-and-groove edges, applied either horizontally or diagonally from the sill to the top plate. Horizontal application is easier, and there is less lumber waste than with diagonal application. But applying the sheathing at a 45° angle makes a strong connection between the floor and wall framing and adds greatly to the rigidity of the structure. This method is often required in hurricane and other high-wind areas. With either method, joints should always be made over

set of rafters nearest one end of the ridge board, first to the ridge, then to the outer walls. Nail a temporary brace long enough to reach the ridge board to a ceiling joist directly below it. Then level the ridge board, plumb the rafters with the wall, and nail the ridge board to the temporary brace. Now erect the pair of rafters nearest the other end of the ridge board.

Splice the ridge board sections together, using plywood scabs on each side of the splice. Tack the far end of the second section to a temporary brace, making sure that it is level. Then erect the rafter pairs at each end of that section. Fill in the remaining pairs of rafters, checking occa-

sionally to make sure that the ridge board remains straight and level. If all rafters are cut and assembled accurately, the roof should be self-aligning.

Cut and install 1×6 or 2×4 collar beams, joining every third pair of rafters (every 4 feet). Collar beams should be in the upper third of the attic crawl space. The temporary braces may now be removed.

Square a line across the end wall plate directly below the ridge board. This marks the position of the central cripple (or gable) stud. Studs on either side are set 16 inches o.c. Measure and cut the cripple studs individually, and notch the ends so that

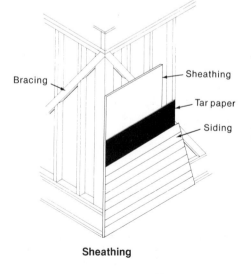

Sheathing

Sheathing is nailed directly to framing; tar paper, then siding cover it.

Plywood wall and roof sheathing in large sheets goes on quickly, covering large areas with little labor.

studs. Boards are cut to fit around window and door openings.

Plywood sheathing is commonly 3/8-inch C-D Interior (it will be covered by finish material and not exposed to the weather). It may be applied either vertically, with the face grain parallel to the studs, or horizontally, with the face grain across the studs. Horizontal application adds extra stiffness to the wall; vertical application makes a better connection between floor and wall framing. (Local building codes may dictate the method.)

Start sheathing application at a corner and work around the structure, with joints occurring over studs or other framing members.

Other panel sheathing materials are applied in the same way. Check manufacturer's recommendations for specific products.

STEP 5
Sheathing the roof
Roof sheathing usually consists of nominal 1-inch tongue-and-groove lumber or 3/8-inch plywood. Lay board sheathing at right angles to the rafters, starting at the lower end of the roof and working up to the ridge. Joints should occur over rafters, and joints of adjoining boards should not be made over the same rafter. Each board should be supported by at least three rafters (32 inches minimum). Use long sheathing boards at roof ends where there are gable overhangs to help attain good framing anchorage.

C-D Interior plywood roof sheathing is laid with the face grain perpendicular to the rafters. End joints are made over a rafter and should be staggered by at least one rafter between every row. A 4-foot (three rafter spaces) offset may be more economical because it eliminates odd-size waste pieces. Make a layout pattern, showing the placement of panels, before you even order the plywood. You may be able to effect a saving of several panels by careful planning of their placement.

Start installing plywood sheathing at a lower corner of the roof and complete the first row before working your way up the roof. Allow spacing of 1/16 inch between panel ends and of 1/8 inch between panel edges. If necessary, cut top row panels to fit up to the ridge.

Complete the framing by removing braces still in place and by cutting away the sole plates in door openings.

Now you have the house fully framed—at least in your mind's eye. But there is a great deal more work to do before you can move in, and this work is beyond the scope of a book on basic carpentry. But if you live in a frame house (and almost everyone does), you should now understand how your house is put together.

Removing or adding interior walls
There are many instances where it may be desirable to remove an interior wall to create a larger room or to erect a wall to subdivide a space. The following instructions tell you how to do both jobs. The instructions for removing a wall are for non-load-bearing walls only. Removing a wall that supports part of your house is a tricky business and should be left to professionals or advanced do-it-yourself carpenters. The instructions for adding walls are designed around the case of an old kitchen that can be successfully divided into a smaller kitchen and a dining room. This case includes the installation of a door. These instructions can be adapted for most partitioning situations throughout the house.

How to remove a wall
Again, these instructions are for non-load-bearing walls only. Before you remove any wall, disconnect any electrical wiring and drain the plumbing pipes in the wall. There is always a possibility of a shock, of a fire, or of water damage to the home. Even if you have shut off the circuit to the plugs in the wall you are removing, there is still a possibility of a live wire in the wall cavity. Shut off power at your circuit box. There could also be a plumbing pipe in the wall even if there are no plumbing fixtures immediately adjacent to it. Determine the path of your plumbing risers from the basement. Drive a long, thin nail through the subflooring from the basement to give you a guide marker

Recommended Nailing Schedule for Wood Framing

Application	Nailing method	Nail size	Number	Placement
built-up girder (three 2X)**	face nail	20d		16″ o.c.,* staggered on both sides
floor joist to sill, girder	toe nail	10d	2	each side
header joist to joist	end nail	16d	3	
header joist, end joist to sill	toe nail	10d		16″ o.c.*
bridging to joist	toe nail	8d	2	each end
subflooring, plywood:				
at edges	face nail	8d		6″ o.c.*
at intermediate joists	face nail	8d		8″ o.c.*
subflooring, boards:				
1×6	face nail	8d	2	to each joist
1×8	face nail	8d	3	to each joist
sole plate to stud (horizontal assembly)	end nail	16d	2	each stud
lower top plate to stud	end nail	16d	2	each stud
stud to sole plate	toe nail	8d	4	
sole plate to joist or blocking	face nail	16d		16″ o.c.*
doubled studs	face nail	10d		16″ o.c.,* staggered
built-up corner studs to blocking	face nail	10d	2	each side
intersecting stud to corner studs	face nail	16d		12″ o.c.*
end stud of intersecting wall to exterior wall stud	face nail	16d		16″ o.c.*
upper top plate to lower top plate	face nail	16d		16″ o.c.*
upper top plate at laps and intersections	face nail	16d	2	
header (two 2X, plus ½″ spacer)	face nail	12d		12″ o.c.,* staggered on both sides
ceiling joist to top wall plate	toe nail	8d	3	
rafter to top plate	toe nail	8d	2	
rafter to ceiling joist	face nail	10d	4	staggered
rafter to valley or hip rafter	toe nail	10d	3	
ridge board to rafter	end nail	12d	3	
rafter to ridge board	toe nail	8d	4	
	face nail	10d	1	top edge
collar beam to rafter				
(2X);	face nail	10d	2	each end
(1X)	face nail	8d	3	each end
wall sheathing, ³/8″ plywood:				
at edges	face nail	6d		6″ o.c.*
at intermediate studs	face nail	6d		12″ o.c.*
wall sheathing, ½″ plywood:				
at edges	face nail	8d		6″ o.c.*
at intermediate studs	face nail	8d		12″ o.c.*
wall sheathing, 1×6 or 1×8 boards:				
horizontal	face nail	8d	2	to each stud
diagonal	face nail	8d	3	to each stud
roof sheathing, ³/8″ plywood:				
at edges	face nail	6d		6″ o.c.*
at intermediate rafters	face nail	6d		12″ o.c.*
roof sheathing, ½″ plywood:				
at edges	face nail	8d		6″ o.c.*
at intermediate rafters	face nail	8d		12″ o.c.*
roof sheathing, 1×6 or 1×8 boards	face nail	8d	2	to each rafter

* On-center (center-to-center) spacing.
** 2X indicates board may be 2×4, 2×6, etc.

if you are not sure which wall the risers pass through.

STEP 1
Identifying bearing walls

Most interior walls are just room dividers that can be removed without structural problems. However, some may be load-carrying or *bearing* walls. Usually, a bearing wall will run the length of the house, roughly in the center. However, it can jog, and its function may not be obvious. The bearing wall helps cut down the span of the ceiling joists, allowing use of shorter lengths of lumber.

Depending on the size of your house, or the type of soil under it, you may have more than one interior bearing wall, or you may have none at all. This can be determined by looking at the construction plans for your home, if they are available.

You may also be able to tell by looking in the attic. An interior wall that runs perpendicular to the ceiling joists (joined by either lapped joints or butt joints) is a bearing wall. Interior walls running parallel to the joists rarely are bearing walls. Homes framed with roof trusses normally have no interior bearing walls.

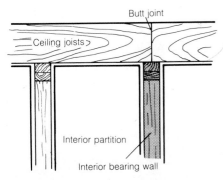

An interior partition does not support joists; bearing wall does.

STEP 2
Removing the wall surface

Gently pry off the base molding. Use a pry bar and hammer for this, but take it easy as you pry. When the molding has been removed, pull out the nails and store the molding out of your way.

If the wall is covered with gypsum board (sheetrock), use a hammer to break out the gypsum board panel back to the studs to which the sides

of the panel have been nailed. If you go very slowly, you can remove the entire damaged panel by removing chunks of the panel and pulling the nails as you go. Do it in small pieces rather than jerking off the whole panel at once. When you come to the panel joints, cut the gypsum board tape that spans the joints along the sides and at the ceiling line with a razor knife.

With the panel removed, you should have a neat, clean hole in the wall with the studs exposed and gypsum board panels overlapping the side studs by about half the width. Repeat for each panel.

If the walls are paneled, work carefully and try to remove the panels intact. To remove the paneling, pry a little at a time along one edge. Do not try to pry one nail up completely before moving to the next. Work on each nail a little at a time until you have freed an edge. Then work on the next set of nails. You may need a long crowbar to reach the middle portion of a wide panel.

STEP 3
Removing the wall framing

The stud wall can be disassembled easily and taken down once the surface material is removed. Use a nail claw to lift out large nails; work carefully to avoid damaging the framing lumber. Use a file to sharpen the claws of the hammer from time to time to keep an edge that will slip under the nail head.

Remove the framing one piece at a time, beginning with the studs. Pull the toe nails out at the top and bottom and "wiggle" the stud out from the remaining framing. Do not hit the stud hard; this could cause damage. If the stud does not come out easily, tap it lightly with a hammer, alternately at the top and bottom. Do not strike the stud in the middle. Continue until all the studs are removed.

To prevent the top plate from falling once all the studs are out, leave a stud at each end to support the top plate (which should not have been nailed to the ceiling joists). Pull any nails out of the top plates and remove it. Then take out the two end studs. Finally, remove the nails or pins holding the bottom plate down and lift it out of place.

How to add a wall

There are times, of course, when a new wall is needed. For instance, consider an older home with a large, 16- × 20-foot country kitchen. If you need the existing dining room as a study/studio but still require a formal dining room for entertaining, the old kitchen could be divided to provide an 11- × 16-foot kitchen, a 9- × 13-foot dining room, and a hallway, as shown. The resulting kitchen can be more efficient and more attractive.

To discuss building a new wall, we will use the example kitchen below. This project will cover all the factors you will need to know. The information can be adapted to build one

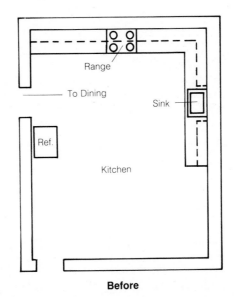

Before

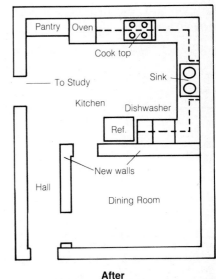

After

straight wall or a more complicated job of building several new walls.

STEP 1
Finding the best location

If you have a solid floor with a good finish, you may build your new wall directly on the existing floor.

Ideally, the wall should sit directly over a joist if located parallel to the joists. If not, you will have to install nailers between joists for secure nailing of the bottom plate. Check the location of studs in the walls into which the new partition will run. If possible, locate the new partition so that it will fit halfway between studs for less opening and patching of walls.

STEP 2
Marking guidelines

Snap two chalklines 3½ inches apart at the proposed location. The 3½-inch width is the same as the actual width of the 2×4 to be used as the base plate of the wall.

Check to see that the dimensions of the two spaces on either side of the wall are what you expected. Arrange furniture or chalk furniture plans on the floor and walk around to appreciate the space. If you feel one side is cramped, adjust the chalklines; once in place, the wall will be permanent and you will have no such choice.

STEP 3
Connecting to wall framing

Open the walls into which the new partition will run. If the junction is very close to a stud, open the wall up

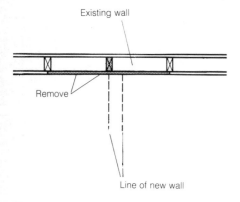

Remove wallboard or plaster to center of studs on either side of new junction.

to the studs on either side of that stud. If the new partition meets the old wall halfway between the studs, open the wall to the centerlines of the studs on either side.

STEP 4
Fastening to floor

If the new wall sits directly over a joist running parallel to the new partition, or if the new partition runs perpendicular to the joists, you will not have to add floor supports for nailing. However, if the new partition sits between joists, you will have to add 2×4 blocking between the joists of the basement ceiling to provide a secure nailing surface for the bottom plate. Nail blocking through the sides of the joists into the ends of the 2×4 sections. Position nailer blocking snugly against the subflooring, working from the basement.

STEP 5
Installing bottom plates and wall "Ts"

Lay 2×4 bottom plate(s) on the floor. Remeasure and check locations before nailing to floor and joists (or blocking) with 12d common nails.

STEP 6
Installing wall "Ts"

The new wall is joined to the existing walls with "Ts." Construct the Ts from 2×4s. Two of the boards in the Ts will fit between the bottom and top plates of the existing wall, as if they were regular studs; fasten them with nails to short sections of 2×4

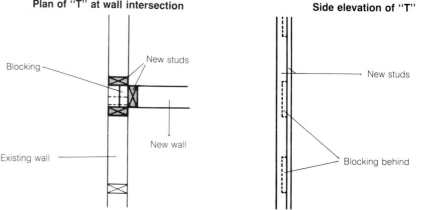

Plan of "T" at wall intersection

Side elevation of "T"

Remove stud that will interfere with installation of T intersection. Blocking holds spacing of the T section that fits into old wall.

blocking. This blocking is also the nailing surface for a third board that serves as the end stud of the new wall. Nail it flush with the blocking, outside the old wall. This part of the T will sit on the new bottom plate and reach from the new wall bottom plate to the top plate. Cut the board 1½ inches short of the distance from the bottom plate to the ceiling to allow for the installation of the top plate. Construct a T for each end of the new wall. Toe nail into existing top and bottom plates and into the new bottom plate.

STEP 7
Adding top plates and studs

If the length of the top plate is short enough so that a single 2×4 will reach from one side of the room to the other, slip the top plate over the Ts and toe nail it into place. Locate studs at 16 inches on center (measure 16 inches from the face on a T

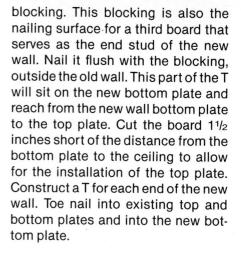

and mark the bottom plate in 16-inch increments); place an "X" where each stud will go. Toe nail the studs over the X marks. If the top plate is in position, toe nail the studs to the top plate. If the top plate is made up of

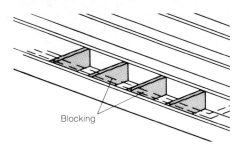

A new partition that runs parallel with ceiling joists requires the addition of nailer blocks between joists. Install these 16 inches on center.

two sections, install a stud at the end of the first top plate section. Then place another stud next to it to support the end of the second section of the top plate.

If you have had to provide special blocking for the bottom plate, you also will have to open the ceiling to provide blocking for nailing the top plate. This will require patching (this can be done when repairing the walls). You will have to cut out the ceiling to expose all of the two joists between which you install blocking. The ceiling opening should reach to the centerlines of the joists. This assumes a two-story house. If you have a one-story house, blocking may be added from above, working in the attic for access to the ceiling joists.

Install the upper wallboard sections first. Rub colored chalk on the junction boxes of any outlets or switches you have installed and hold the wallboard against the boxes. This will mark the position on the sheet. Cut out with a keyhole saw.

STEP 8
Framing a door

Decide on the size and type of door you want. To simplify the process, buy a prehung door; this unit comes with installation instructions and precise measurements for the rough opening. Before you begin wall framing, mark the rough opening width for the door on the bottom plate. Place a stud on either side of the opening so that the space between the studs is 3½ inches greater than the rough opening of the door. This leaves room for studs and shims. Cut out the bottom plate between these studs.

STEP 9
Installing jack and cripple studs

These are 2 × 4s that are nailed to the previously installed studs. The jack studs are as long as the rough opening is high. Cut two 2 × 4s for a doubled header to fit across the jack studs. Complete the framing by installing cripple studs between the top plate and the headers, as shown. If you substitute an extra-wide header for the double 2 × 4s, you will not need cripple studs.

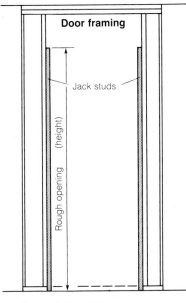

Door framing

Jack studs

Rough opening (height)

Remove bottom plate

STEP 10
Covering the wall

If you are planning to add any electrical outlets or switches, do the electrical work at this time. When the framing and any other work is complete, cover the wall with plasterboard. Begin by patching any holes made for the installation of the Ts and blocking. Cut the wallboard to fit and nail it in place with as narrow seams as possible. Set the nails just below the surface of the wallboard, dimpling the surface with the hammer. The seams and dimples will be filled in later.

For the expanse of the wall, install the wallboard (½ or ⅝ inch thick) horizontally. This places the long seam at a comfortable working level and will make finishing easier. Mark the doorway on the wallboard, then

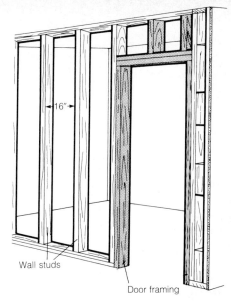

16"

Wall studs

Door framing

cover the wall (including the doorway) completely. Cut out an opening for the door flush with the studs you installed on each side of the doorway by drilling a hole near the center of the door area and, using a keyhole or saber saw, cutting over to the studs. Cut out the opening.

STEP 11
Finishing the wallboard

Apply a thin coat of joint compound on all the seams and over all the nail heads. Then, using a "finishing" knife (similar to a putty knife), apply a smooth layer of seaming tape. When this is dry, apply a second coat, about 6 to 8 inches wide. Let this dry and sand smooth. Apply another coat that is 12 to 14 inches wide. When this has dried, wipe with a damp sponge to smooth. Sand if necessary.

STEP 12
Finishing the installation

Hang the door in the opening as directed by the manufacturer. This usually requires removing the door itself from the hinges and slipping the interior frame into the rough opening. Level and plumb the frame with shims as needed and attach to the wood framing. Rehang the door on the hinges and install facing materials as provided or as directed.

Install the moldings, baseboard, and outlet face plates. Then complete finishing, for example, a final coat of paint.

9 SANDING AND ABRASIVES

The work of all but the roughest projects is not finished when the construction is done. You want your project to be pleasing to the eye, so some sort of finish is required. And before you apply a finish on most projects, you have to smooth the surface by sanding; a satinlike surface on a project is as much the product of good sanding as it is of the finish applied.

Ready-made sanding blocks.

Types of sandpaper

The term *sandpaper* is used to describe a variety of sheet abrasives. Among the most common are flint, garnet, emery, aluminum oxide, and silicon carbide. They may be mounted on paper or cloth, in "open-coat" or "closed-coat" density. The following are capsule descriptions of the common types of sandpaper.

Flint. The least expensive sandpaper is flint paper (flint is a gray mineral). This paper wears down quickly and is used mainly in removing paint. Because paint will clog the grit on the paper and reduce its usefulness before the paper is worn smooth, pieces have to be discarded frequently; it is advisable to use as inexpensive a paper as possible for paint removal.

Garnet. The grit on garnet sandpaper is much harder than on flint paper and is more suitable for use in woodworking.

Emery. This abrasive is recognizable by its distinctive black color. This is still fairly widely used as a metal abrasive, although more recently developed abrasives may be more effective.

Aluminum oxide. This is probably the most popular sandpaper abrasive for fine woodworking and furniture finishing. It is also the best choice for power sanding. The oxide is reddish in color, and its grit is very sharp and much harder than that of the first three papers listed here. While it is popular for use on wood, it can also be used on metal. Aluminum oxide is more expensive than other abrasive paper, but it lasts so long that it is often cheaper to use in the long run.

Silicon carbide. This is the hardest, sharpest sandpaper of all. The bluish-black material cuts extremely well and is commonly used for such tough jobs as finishing metal or glass and for floor sanding.

Sandpaper classifications

Apart from the type of material that gives sandpaper its abrasive quality, there are other characteristics that affect the performance of abrasives. There is a paper for any job from the roughest stripping to the most delicate finishing. The following are the basic classifications of sandpaper.

Grit. The *grit*, the particles that are adhered to the backing material, are each identified by a number of labeling systems. There is a Retail System that carries word descriptions (coarse, medium, fine). The Old System identified papers by numbers that run from 10/0 to 4 to reflect increasing coarseness. The currently used Industrial System identifies grit with numbers from 600 down through 16. In the Industrial System, the grit specified is coarser as the numbers become smaller. (See the *Abrasive Papers* chart.)

Density. Another factor that affects the way sandpaper functions is the *density* of the grit—how close together the granules are on the backing. There are two classifications: closed-coat and open-coat. *Closed-coat* indicates that the grit material blankets 100 percent of the surface; *open-coat* indicates that the grit covers from 50 percent to 70 percent of the surface. Open-coat may not look very "open," especially when the abrasive is of a fine density.

Generally, closed-coat sandpapers are designed for fine finishing. Closed-coat paper tends to clog quickly. Wood particles are caught and fill the spaces between the pieces of grit. Because open-coat sandpapers do not clog as easily they are a better choice when you do a first sanding to remove a lot of material from the wood. The usable life and cutting action of any sandpaper can be extended and improved by rapping the paper on a hard surface from time to time to dislodge wood particles, or by cleaning it with an old toothbrush.

Adhesives. The grit is held to the backing on sandpaper by any one of a variety of glues.

Hide glue is used on sandpapers intended for light to medium work. Hide glue is not waterproof, so the paper cannot be used on a wet surface. Thermo-setting resin is used to secure grit on papers where the work is harder, such as floor sanding. Waterproof resin bonds grit to backings that are waterproof. Such glue allows the paper to be used with oil or water for extra fine and smooth finishing.

Weight and backing. The backing of the sandpaper may range from paper or cloth to a combination of cloth and paper or even a plastic material. The paper comes in various weights, which lend certain advantages or disadvantages.

A weight paper is the lightest of all and the first to wear through. Use it only for lightly touching up wood.

Moving up the scale, *C* weight is stiffer and stronger than *A* weight and is used for coarse machine sanding. *D* weight is even stronger and is for heavy machine sanding. *E* weight is the strongest of all. It is designed to be used with floor sanding machines or belt sanders.

In the cloth backings, the weights are *J* weight and *X* weight. *J* is the lighter of the two and is used on curves and other shapes. The *X* weight cloth is designed for coarse grit abrasives used with a belt sander.

Fiber weight backings are very heavy and are not needed by the do-it-yourselfer for any of the projects discussed in this book.

Other abrasives

There are two powdered stone abrasives commonly used in fine finishing. These abrasives are applied with an oil. Depending on the situation, you may mix the powdered stone with lubricating oil and apply with a rag, or you may dip the rag in the oil and then in the stone before applying.

Pumice. This is a lava. The stone is porous and relatively soft. When pumice is crushed to a powder, the stone is a medium to fine abrasive.

Rottenstone. This is decomposed limestone, a form of sedimentary rock. This material crushes to a very fine abrasive powder. Applied with oil, rottenstone is used to create a smooth finish to projects like furniture.

Either of these abrasives may be used to create the glassy finish of a French polish.

Steel wool

Steel wool is also used for smoothing wood. It comes, like sandpaper, in a variety of grades. Each has a number and name description. Following is a brief description of each grade and its common uses.

0000. This is the finest grade of steel wool usually available. It creates a satin-smooth finish on fine woods. It is used for such jobs as rubbing down shellacs, lacquers, varnishes, waxes, and oils. It is also used for cleaning delicate instruments.

000 extra fine. This is used for both cabinet work and auto finishing as well as to remove minor cracks or checks and burn marks in a finish. This grade performs well in removing paint spots and splatters. It is also the grade used to polish metals.

00 fine. This grade is used to cut gloss finishes of paint to a semi-gloss. It is also desirable for cleaning and polishing wood floors, plastic tile, and terrazzo. A varnish remover is applied with 00 fine steel wool to remove old finishes.

0 medium/fine. This grade is not widely used in woodworking. It is primarily used for cleaning aluminum, copper, brass, or zinc and many metal objects such as barbecues, pots, and pans and for removing rust.

1 medium. This grade is used to prepare wood for a first coat of paint. It is also used with soap and water to clean various flooring materials such

Abrasive Papers

Grade (Industrial system)	Grit (Old system)	Description (Retail system)	Use
16	4	very coarse	very rough work; unplaned
20	3½	very coarse	wood; initial machine sanding
24	3	very coarse	of floors
30	2½	coarse	initial sanding when necessary
36	2	coarse	
40	1½	coarse	
50	1	coarse	
60	½	medium	intermediate sanding,
80	1/0	medium	especially of softwoods
100	2/0	medium	
120	3/0	fine	preparatory sanding of hardwoods;
150	4/0	fine	final smoothing, especially of
180	5/0	fine	softwoods
220	6/0	very fine	final sanding; sanding
240	7/0	very fine	between finish coats
280	8/0	very fine	
320	9/0	extra fine	sanding between finish coats;
360		extra fine	smoothing the final coat of finish
400	10/0	extra fine	
500		super fine	sanding metal, plastic,
600		super fine	ceramics

as rubber, asphalt, linoleum, and resilient floors.

2 medium coarse. This grade is used to clean rust and dirt from garden tools, glass, brick, metal, and stone. It is not often used on wood.

3 coarse. This grade is used to remove old paint and varnish. After the remover has caused the paint or varnish to liquify, 3 coarse steel wool can be rubbed on the surface to loosen the remaining paint.

Sanding by hand

For most woodworking projects hand sanding is the technique of choice because the results can be controlled better. Also, hand sanding is unquestionably safer than machine sanding.

The object of sanding is to smooth the wood. The final character of the smooth surface will be dictated by the finish chosen so there is some variation in final smoothness from job to job. If you plan to paint the piece with enamel, you need not sand as smoothly as you would if you were going to use a clear finish. The enamel finish will be glasslike as long as the sanding is reasonably smooth and the paint is applied with careful brushwork. A clear finish requires an extremely smooth surface for a high-quality surface.

To soften sandpaper, pull the back side over a corner.

Using a sanding block

You will undoubtedly get a smoother finish and find the job easier if you use a *sanding block* when you sand a flat surface. There are commercially made sanding blocks available, but it is easier to make your own.

Cut sandpaper to fit the wood block you will use.

Fold sandpaper tightly around the block and secure with thumbtacks.

If you are sanding a curved surface, glue felt or foam rubber around a section of dowel or a broomstick and wrap sandpaper over it. This tool will help you smooth inward curves.

Power sanding

Although hand sanding is of primary concern to the fine woodworker, there are some jobs that may be done by machine. There are basically three kinds of sanding machines: the disc, the belt, and the finishing sander (see *Portable Power Tools*). The first two have very limited use, if any at all, for the wood finisher.

The logical machine sander for the fine woodworker is the finishing sander. This machine uses a rectangular pad and precut pieces of sandpaper. The action of the sander is either straight line (back and forth) or orbital. Most finishing sanders are orbital; however, some come with both actions. A combination pattern sander has a lever that allows you to use either straight-line or orbital sanding action.

All of these machines, like most portable power tools, are available with all-plastic housings; they are double-insulated to guard against electrical shock. Many models also come with dustbags and vacuum action. This feature reduces the irritation and mess of sawdust in the air and around the shop.

The straight-line action of a finishing sander is slow, but it produces a smoother finish than the orbital action. Although the orbital does go against the grain in its round-and-round action, many workers end up satisfied with the smoothness of the finish it produces.

The secret of using a finishing sander is not to press down on the machine but to let the weight of the unit do the work. Pressing inhibits the sanding action of the machine; restriction of the movement may damage the machine motor.

If you are sanding vertical trim, the weight of the machine will not be exerted on the surface. In this case you will have to apply a little pressure, but don't exert too much.

When using a finishing sander— or a belt sander—you will use various sandpapers, starting with a relatively coarse grit and ending with a fine paper to produce the smooth finish you want.

Basic sanding procedures

You should always start a job with the smoothest usable grade of abrasive material. If you start with material that is coarser than necessary, you will cut small grooves in the surface. These will have to be removed in subsequent sandings. This means more work for you and less chance of a complete success in your finishing project.

For final power sanding, use a reciprocating sander with the grain.

Achieving a smooth finish

To prepare bare wood, follow the steps given below in order to achieve a fine, smooth surface suitable for application of a clear finish.

STEP 1
Using progressively smoother papers
Start sanding with a 220 or 280 paper. If the wood is a softwood, use a 220 paper. If you are finish sanding a hardwood, use the 280 or finer papers. Always sand *with* the grain. Wipe the dust off the surface regularly, and use a *tack rag* (a cloth pretreated with a gummy material) to get the wood clean of all loose dust and grit. Keep sanding until the surface is as smooth as you wish. Run your fingers over the surface to check the progress of the work. If you intend to paint the surface with enamel, the smoothness achieved with a 150 paper should be good enough.

STEP 2
Raising the grain
At some point, it will seem that you have sanded the wood as smooth as possible. At this point you must *raise the grain*. First, wipe away all dust with a dry, lint-free cloth. Then soak a clean rag in water, wring it out, and dampen the sanded surfaces. Let the wood dry at least twenty-four hours. Run your hand over the wood. If you feel a fine fuzz, the water has expanded the wood and the grain has been raised. Rub the surface gently with a fine sandpaper to take off the "whiskers." Repeat this procedure and, if necessary, keep repeating it until no grain is raised.

STEP 3
Using a sanding sealer
You may achieve an even smoother finish by using a *sanding sealer*. These are available in paint stores and come with instructions for application. Such sealers are also used over filler material so that after wood is filled, application of a final clear finish does not lift the filler. It is also used over raw wood to prevent excessive absorption of stain.

The sanding sealer you buy in the store is not cheap. You can make your own by mixing one part clear shellac with four parts denatured alcohol. Brush it on. When the mixture dries, do the final finish sanding.

Wood fillers
Open-grained woods are often treated with wood filler to create a super-smooth surface before finishing. Oak, ash, walnut, beech, elm, teak, and rosewood may be so open-grained and porous that filling is mandatory for a satisfactory finish. Fine-grained woods—pine, fir, cedar, maple, redwood, cypress —are seldom filled. Some craftspeople feel that the use of fillers detracts from the natural look of the wood. The choice is yours.

Wood fillers are available in paste and liquid form. Paste fillers are composed of inert stones dissolved in pure linseed oil and are best for open-grained woods. They are available in almost any wood tone and can be mixed with oil colors to match the wood to which they are being applied exactly. Read the label for thinning instructions. Usually, the paste is thinned with turpentine, benzine, or paint thinner to the consistency of heavy cream—just enough so that it can be brushed on easily. The surface to be filled should be wiped clean with a lint-free cloth and lightly dampened with the thinner. Apply the filler with a brush, working it *along* the grain so that every wood pore is penetrated.

When the entire surface has been covered, allow the filler to almost dry on the surface, usually about ten to fifteen minutes—the surface should appear cloudy. With a pad of burlap, rub the surface *across* the grain to force the filler firmly into the wood and remove any excess. When all traces of the filler have disappeared from the surface, use a clean, soft cloth for a final rubbing along the grain. Let the filler dry for twenty-four to forty-eight hours, then sand lightly with very fine abrasive paper (220 or 240). Dust the surface thoroughly.

Liquid fillers are used to fill fine-grained woods. These usually have a base of varnish or lacquer, with relatively little solid matter. The filler is brushed along the grain onto the wood surface. After it dries completely, the surface is sanded with fine (150 or 180), then very fine (220 or 240) abrasive paper. If some wood pores remain unfilled, apply a second coat of filler, allow it to dry, and sand smooth.

10 FINISHING

The most skillfully crafted project will not be pleasing if it is poorly finished. This aspect of woodworking requires just as much careful attention as the cutting and joining of the wood parts, and perhaps even more patience. Don't rush it, and always follow the manufacturer's directions for the finish that you are applying.

Wood stains

Many furniture woods look their best when left *natural*, that is, with only a clear protective coating applied over the surface. Such a coating will almost surely deepen the wood tone somewhat, making it appear slightly darker than the raw wood. Such species as maple, oak, cherry, cedar, and rosewood are usually given a natural finish. On the other hand, ash, beech, poplar, and pine are often stained to emphasize attractive grain patterns or other features. Stain may also be used to make a new piece match existing furniture, or even to blend better with a room's decor.

Experimentation is strongly advised before staining. Apply the stain and the intended finish to a scrap of the same wood species, or to an underside or inconspicuous corner of the piece, to see what the result will be. Such precautions may slow you down a bit, but it's a lot easier than trying to change the look of the entire piece later on. The following are common stains.

Wiping stains. These are stains that contain pigments in a carrier (usually linseed oil); they are absorbed by the wood fibers and emphasize the grain. Wiping stains are easy to use, especially on most softwoods. Some very tight-grained woods resist soaking up the stain.

Wiping stain colors run the gamut from handsome ash to rich, deep walnut, but the final shade may vary according to the wood species being treated. That is why it is important to experiment first. A little common sense helps too. You don't have to be an expert to figure out that a fruit-wood stain (fairly dark) will darken beech (light) but will have little effect on walnut (also dark).

Penetrating stains. These usually consist of synthetic resins dissolved in benzine or turpentine. The resins penetrate the wood fibers and harden them against further penetration, locking in the color. But again, make sure that it is the color you want—once the penetrating stain is

Wiping stain is one of the easiest finishes to apply. Put it on with a lint-free cloth, then wipe it off: do not leave surface wet to pick up dust.

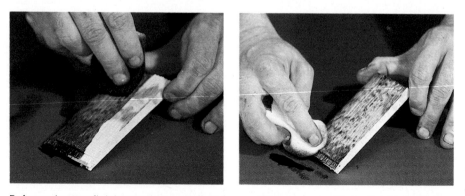

Before using any finish, test it on a piece of scrap of the same wood to be finished so you will know what the final product will look like.

applied, it's not easy to reverse the result. Penetrating stains are also available in a wide range of wood tones.

Non-grain-raising (NGR) stains. As the name implies, NGR stains do not raise the wood grain (neither, by the way, do oil-based stains). NGR stains are generally made of aniline or synthetic dyes. They are purchased in powder form and are mixed with alcohol, usually one packet of powder to one quart of denatured alcohol. NGR stains give the wood a brilliant color; darker tones, such as walnut or cherry, may be so intense that twice as much—or even more—alcohol must be added. Experiment on scrap until you achieve the desired result. Alcohol evaporates quickly, so a second coat of stain may be applied right after the first. But you must work fast when applying the stain; slow application leads to uneven coverage. Always apply too little rather than too much; if the surface is too light, you can simply go over it again.

Water-based stains. These are similar to alcohol-based stains. The difference is the water, which *will* raise wood grain. For this reason, water-based stains are less desirable than other stains. Wood treated with these stains requires extensive sanding before it is ready for the final finish, and this sanding often affects the color of the stained surface.

Using stains

Whatever type of stain you use, make sure you read the label and follow the instructions to the letter. You can't blame the manufacturer for unsatisfactory results if you haven't followed instructions.

After the work surface has been smoothed, wipe it with a clean cloth. Even better, vacuum it. Many stains are applied with a soft cloth; others, with a brush. Rub or brush the stain evenly over the surface, both across and with the wood grain. Allow to dry according to label directions, then wipe away the excess with a clean, soft cloth, first across the grain, then with the grain.

If the shade is too light, apply another coat of stain in the same manner. If it is too dark, remove some of the stain by rubbing with its solvent (turpentine, benzine) as quickly as possible.

NGR stains should be brushed

Early American	Pine	Cherry	Birch	Fruitwood	Oak
Puritan Pine	Pine	Special Walnut	Birch	Ebony	Oak
Fruitwood	Pine	Puritan Pine	Birch	Dark Walnut	Oak
Provincial	Pine	Colonial Maple	Birch	Driftwood	Fir

Wood stains applied with a soft cloth will darken wood to desired tone.

onto the surface with long, straight strokes. Since the alcohol evaporates quickly, try to avoid overlapping; the overlaps may appear darker than the rest of the surface, and you'll end up with streaks. If this does happen, go over the entire surface again to even the color.

When the desired tone is attained, and the stain has dried thoroughly, sand the surface lightly with a very fine abrasive paper. Worn paper is good for this job, since the purpose is simply to remove dust particles from the surface. Alcohol-based stains should be given a coat of equal amounts of shellac and alcohol after sanding. Since these stains tend to "bleed," a hard varnish finish is recommended.

Sealers

The purpose of a *sealer* is to lock in the stain so that it does not react with the finish coat. (Do not confuse this type of sealer with a penetrating sealer, which is itself a finish to be discussed later.) The usual sealer is a thinned version of whatever finishing material will be used: for varnish, half varnish and half turpentine; for lacquer, half lacquer and half lacquer thinner; for shellac, half shellac and half denatured alcohol. Actually, the shellac-alcohol sealer will work under any finish.

Sanding sealers may also be purchased at paint supply stores. These are basically mixtures of shellac and varnish, along with some other ingredients. These work effectively and dry quickly, although no more so than other sealers.

Sealers should be spread on the work surface with a brush, in the direction of the grain. After the material has dried completely, sand the surface lightly with 220 or 240 abrasive paper; do not apply so much

Ipswich Pine	Pine	Natural	Birch	Jacobean	Oak
Special Walnut	Pine	Fruitwood	Birch	Natural	Oak
Colonial Maple	Pine	Red Mahogany	Birch	Golden Oak	Oak
Conestoga White	Pine	Early American	Birch	Provincial	Oak

Common Wood Finishes

There are many finishing products on the market, some very new and others very old. No one finish is the "best" for all purposes. Most woodworkers develop strong preferences and prejudices in favor of one type of finish or product. We suggest that you try several, then form your own opinions about your favorites.

Finish	Description	The good points	The bad points	Recommended use
shellac	lac dissolved in alcohol: orange shellac darkens wood; white shellac adds only slight tone	inexpensive; easy to apply; quick-drying	brittle; low resistance to water, alcohol; limited shelf life	furniture that will not be exposed to abuse; as undercoat for varnish
varnish	resins in an oil carrier; comes in gloss, semi-gloss, low-luster finishes	hard, clear, durable, smooth; resists water, acids, alkalies, alcohol	fairly difficult application; slow drying; dust will adhere if care is not taken	general furniture use
polyurethane varnish	similar to varnish in appearance, but chemically different	easy to apply; quick drying; harder and more durable than varnish	more expensive than other finishes; may be incompatible with other materials (check label)	general furniture use; outdoor projects
lacquer	resins in thinner; available in a range of colors in flat or gloss finish	hard, durable; quick-drying	difficult to apply without special spraying equipment	where extra-hard finish is desired
penetrating sealer	resin finish that penetrates and hardens the wood without covering	easy to apply; provides excellent protection with "natural" look	expensive; may be incompatible with other finishing materials (check label)	where wood must be protected, but a hard, glossy finish is undesirable
linseed oil	boiled linseed oil, thinned and rubbed into wood pores	gives wood a rich, soft luster; somewhat resistant to heat, scratching, less so to water	tedious application; all surfaces, including undersides, must be treated; finish must be renewed every six months	where a dark, antique appearance is desired; where a hard surface is not desired
French polish	traditional technique using shellac and linseed oil	fairly durable; gives wood a rich, warm glow	fairly difficult and tedious to apply; requiring several rubbings	where an antique appearance is desired
wax	not a finish; used to shine and protect finishes	inexpensive; easy to apply	requires frequent buffing and renewal	over hard finishes that get frequent care
enamel	varnish with pigments added; comes in gloss, semi-gloss, flat finishes	hard, durable, smooth, opaque; resistant to moisture	fairly difficult to apply	where an opaque finish is desired for appearance or to conceal wood grain

pressure that the abrasive cuts through the sealer. Continue sanding until the surface is as smooth as you can get it—this is likely to be the final rubdown before application of the finish.

After sanding, wipe the surface with a damp (not wet) cloth to remove all dust. Then wipe with a clean, dry, lint-free cloth.

Shellac

Shellac as a furniture finish has a long history—it was used in India at least 2,000 years ago. Shellac is a natural product, secreted by the lac-cifer lacca (lac) bug that inhabits parts of India and Ceylon. The secretions (and the bugs themselves, which become covered with it and die) are harvested, crushed, ground, and processed into flakes. The flakes are then dissolved in alcohol.

Shellac is easy to apply and dries quickly to a beautiful finish, especially on light woods. Unfortunately, the finish is somewhat brittle; it scratches easily and is adversely affected by liquids, especially alcohol. It is often used as a first coat under another finish, such as varnish. When using shellac, always buy a fresh supply. It deteriorates chemically after manufacture and has a short shelf life, usually about a year at most. After that, it refuses to dry properly, and the surface remains sticky.

Applying shellac

Shellac should be applied in a warm, dry, dust-free room. Just before application, go over the surface one final time with a clean, soft cloth. Dip a clean, natural-bristle brush into the shellac, load about one third of the

A piece of plywood without finish (above) and after shellac has been applied (below).

bristles, then apply it to the surface, working along the grain with long, even strokes that overlap slightly. After the first stroke, *tip off* with a second stroke, running the very tip of the brush along the previous stroke to smooth it right to the edge of the work. Do not brush back and forth over the same area, and work quickly. When the entire surface has been covered evenly, stop. If there are brush strokes showing, don't attempt to go over them. They will be covered by the second coat.

Allow the shellac to dry completely (check label directions), then rub it with 240 or 280 abrasive paper, working in the direction of the grain. Wipe with a clean, dry cloth and apply a second coat. Subsequent coats may take longer to dry; otherwise, the procedure is the same as for the first coat.

Varnish

Varnish provides a tough, durable finish that is resistant to water, acids, alkalies, and alcohol. It is available in gloss, semi-gloss, and low-luster finishes. It flows on easily, but is slower drying than shellac or lacquer. Quick-drying types are available and recommended.

Varnish is made of resins or gums (the "body"), linseed oil or tung oil (the "vehicle"), drying agents, and a thinner, usually turpentine. "Spar varnish" contains additives that make it impervious to salt water, but it also includes only a small amount of the drying agent so that it doesn't become sun-damaged. This makes it fine for use on boats and some other outdoor applications like wooden sleds, but it is not recommended for general use.

Spar varnish (left) and polyurethane (right) are both clear finishes. Test differences on wood scrap before use.

Polyurethane varnish is chemically different from standard varnish. It produces a similar appearance but provides a considerably harder and more durable surface, almost like a sheet of clear, very hard plastic. In addition to the abuses withstood by other varnishes, it resists chipping, cracking, peeling, and scuff marks. Polyurethane varnish costs slightly more than other finishes, but it is well worth it in most cases. A note of caution: some polyurethanes cannot be used over shellac, lacquer, or filler; check the label for applicability before buying finishing materials.

Lacquer

Lacquer has been used as a finishing material almost as long as shellac has. Early Oriental artisans made it from a plant similar to poison ivy. Some exquisite Japanese and Chinese furniture pieces were given as many as 300 coats of lacquer and have retained their beauty for centuries.

Modern lacquers are formulated of natural or synthetic gums or resins, nitrocellulose, softeners, solvents, and a thinning agent. They are used primarily in furniture factories, where they can be sprayed on quickly and cheaply. Brushing lacquers are also available in both clear and a range of colors, and in flat or gloss finish. They dry quickly with good resistance to wear but not to moisture. They also have a tendency to cloud, to form pinholes, and sometimes to bleed. The rapid drying makes them difficult to apply, and brush marks on the surface are hard to avoid.

Penetrating sealer

Penetrating sealers (not to be confused with penetrating stains or ordinary sealers) produce a finish that is resistant to moisture, alcohol, acids, alkalies, heat, cold, burns, scratches, grease, and abrasion. They are easy to apply, even for a beginner. The sealer sinks into the wood pores, filling cavities and becoming part of the wood itself. For this reason, it is very difficult to remove, so make sure you will like the result before using a penetrating sealer. The "natural" look it imparts may not work on all furniture pieces —much depends on the grain and tone of the wood. The appearance of no finish at all can enhance woods with charm of their own; on other woods, penetrating sealer may look very drab, or worse. Experiment first.

Linseed oil

A classic finish preferred by some traditionalists for antiques and "modern antiques," *linseed oil* produces a rich, lustrous glow on many woods. Its surface does not show minor heat or scratch marks, but it is not very resistant to moisture. Linseed oil darkens the wood (which may or may not be desirable), so experiment first. Linseed oil finishing is also a tedious process, requiring a great deal of time, perhaps over several months (although the piece can be used during the finishing process). Consider all these factors carefully before deciding whether linseed oil is right for your project.

Reducing the Cut of Shellac

The standard cuts (5-pound and 4-pound) of shellac must be diluted before you use them. This table gives you the amounts of alcohol you need to reduce each *quart* of shellac.

To Reduce	To	Add
5-pound cut	3-pound cut	7/8 pint
5-pound cut	2-pound cut	1 quart
5-pound cut	1-pound cut	2 2/3 quarts
4-pound cut	3-pound cut	1/2 pint
4-pound cut	2-pound cut	1 1/2 pints
4-pound cut	1-pound cut	2 quarts

French polish

French polish is a traditional finishing technique, rather than a finishing material. It utilizes shellac and linseed oil to impart a rich, warm glow to a wood surface. But, as with a linseed oil finish, be prepared to expend some effort—it's not a simple matter of brush it on and let it dry.

The shellac should be a 1-pound cut, or thinner if you prefer, in either orange or white, depending on the effect desired. Use only boiled linseed oil; raw oil will leave the surface permanently tacky.

Make a pad of several thicknesses of lint-free cloth. Dip the pad into the linseed oil, then squeeze it to remove most of the oil. Then dip the pad into the shellac.

The secret of French polishing is that you never start or stop the rubbing process while the pad is on the surface of the work. Begin a circular pattern with the pad just off the end of the piece and continue it onto the surface, applying enough pressure to force the shellac into the wood pores but not so much that the pad sticks or stops. (A little practice will show just how much is "enough" but not "too much.") When the pad starts to dry, work your way toward an edge, maintaining an even pressure and continuing the circular motion until the pad is completely off the surface. Dip the pad into the shellac and add a few drops of linseed oil (not more than that—the oil serves only as a lubricant so that the pad doesn't stick). Then work the pad back onto the surface where you left off.

Continue rubbing in a circular pattern, and off and onto the edges for refills, until the entire surface is covered. Allow to dry for twenty-four hours, and repeat the process. Keep repeating it until the wood surface has just the sheen you want. With French polish, the more you work, the better it looks.

Wax

Wax can add protection and impart a rich appearance to some finishes but, because of its low resistance to heat, wax should never be used as a finish for raw wood. Some polyurethane finishes should not be waxed, however; check the label for manufacturer's recommendations.

There are countless waxes on the market. The best are those containing a high percentage of carnauba wax, the hardest type available. Candelilla wax and beeswax are also good. Apply the wax following manufacturer's directions. Brisk buffing is essential to bring out the beauty of the finish. A lambswool bonnet attachment on a 1/4-inch electric drill will make the buffing go much faster and easier.

Enamel

When an opaque finish is desirable for a woodworking project, enamel is usually the choice. Enamel is essentially varnish with pigments added, and it shares varnish's qualities of hardness and durability. It comes in a full range of colors in gloss, semigloss, and flat finishes. *Gloss enamel* is the most durable; *semigloss* has less shine and may be more desirable for some projects. *Flat enamel* is not normally used for furniture, although it may be a good choice for shelving and similar uses.

The rubdown

- Finishing materials and techniques often produce a high-gloss surface appearance that some people find "a bit much." To tone down this gloss and impart a look of age (if that is desired), *rubbing down* may be done after the finish has dried completely.

First, sand the surface lightly with 360 or 400 extra fine abrasive paper, working along the grain. Wipe the surface clean with a lint-free cloth. Mix pumice, for a satin finish, or rottenstone, for a semi-gloss luster, with light machine oil (*not* linseed oil) to a pasty consistency. (Pumice and rottenstone may be purchased at paint stores.) Rub the paste into the surface with a soft cloth. Apply only light pressure—you don't want to rub right through the finish. After rubbing the entire surface, wipe it clean with a lightly dampened cloth. Use a chamois cloth to dry the surface. If the surface is still too glossy for your taste, repeat the procedure starting with the pumice or rottenstone.

Distressing is accomplished by making tiny dents, long scratches, and other marks filled with dark wood paste.

Special effects and finishes

There are many special effects that will give your woodworking projects a unique appearance. Again, you are cautioned to experiment on scrap first.

Many woodworkers choose to make their new pieces look old—instant "antiques." You can accomplish this in various ways.

These special finishes include distressing (damaging the surface to give it a "character"), antiquing (giving the surface a patina of old age with glazes), tortoise shell (using glazes to give the surface a tortoise shell finish), liming (applying a faint overtone of white on wood with a white-pigmented filler), pickling (combinations of stain and sealers with a contrasting color of pigmented filler), and marbleizing (simulating the veins and markings of real marble with color).

Each of these is a process that requires practice and experimentation. You may consult any of several excellent books on the subject for instructions on applying these special finishes. With practice you will be able to imbue your wood projects with any style—ancient to contemporary.

Veneering

Veneering is not a finish as such. Rather, it is a cover-up. It allows you to build a project of less expensive, readily available wood or plywood, then cover it with a very thin piece of a more exotic species, giving the appearance that the entire piece is crafted of Carpathian elm, satinwood, maple, or whatever other wood you choose. Fakery? Well, maybe, but it's fakery with a long and noble history. The Egyptians used veneering as early as 1500 B.C., and many artifacts and furniture pieces found in the ancient tombs (including that of King Tutankhamen) display their mastery of the art. During the golden age of Greek civilization, veneer was frequently employed for decorative effect on furniture, and the early Romans copied the technique and became masters in their own right. Today, with the high cost and scarcity of exotic wood, veneering makes more sense than ever. It also allows you to build projects using stronger (if cheaper) structural woods that may be less attractive, then hiding them beneath a beautiful top layer.

Veneer is a thin (usually $1/28$ of an inch) sheet of wood cut from a log or a segment of a log (*flitch*). It is produced in various ways, each imparting distinctive effects. Sawed veneer is cut in long, narrow strips from flitches specially selected for figure and grain. Either side can be glued, so the most attractive face can be exposed. Sliced veneer, also cut in long strips, is produced by moving a flitch against a heavy knife, shaving off the wood. It is similar to sawed veneer, although the grain patterns may vary. A rotary-cut process turns out continuous sheets of flat-grained veneer by revolving a log against a very sharp knife. Modifications of this process are used to produce highly figured veneers from stumps, burls, and other irregular parts of a log. Rotary-cut veneers have an "open" side and a "closed" side; apply glue to the open side.

Not all lumberyards carry veneers, and those that do may not have a wide selection. Check the Yellow Pages under "Veneer" or "Plywood & Veneer" to see if there is a supplier near you. You can also buy veneers from one of the large woodworking mail-order houses. Veneer is sold by the square foot.

Carpathian elm burl

Sapele

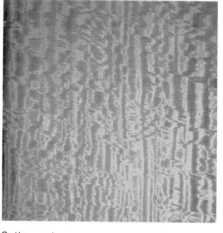

Satinwood

Bird's-eye maple

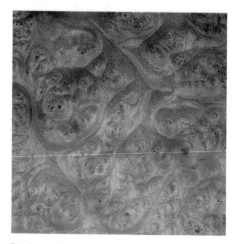

Birch

Poplar burl

WOODWORKING PROJECTS

As you sharpen your woodworking skills, you will probably want to design your own projects—furniture, cabinetry, built-ins, whatever—to suit your specific needs, wants, and taste. The following projects are included to get you started. They range from the very simple to the fairly complex, but none is beyond your capabilities if you have read and practiced the techniques in the preceding chapters.

A few reminders: whenever possible, select your own lumber at the lumberyard. Eye each piece to make sure it is as free of imperfections as you can find, straight, and without warps or crowns. When laying out the work, take your time, and double-check before cutting. Otherwise, you may end up with a lot of waste and a return trip to the lumberyard for more material. Practice "Safety First" always. Make sure that you use the right tool for the right woodworking job, and use it the right way. Keep cutting tools (saws, chisels, planes) sharp, not only to do their jobs properly, but also for safety's sake. Never force a cutting tool—let it do the work. Hold materials securely when you are working on them. The best way is in a vise, if the particular operation allows it. When using adhesives and finishes, follow the manufacturer's directions for surface preparation, application, and drying time. If you don't, you can't blame the product for poor performance or failure.

The lumber table and chair set is handsome and easy to build, offering a refresher course in woodworking basics in the process. You can build one or the other, or both, and finish them however you wish.

A few simple curved cuts make this early American-style shelf an exciting decorative piece. You can use the instructions for this project as inspiration and create your own variations.

Slat benches with planter boxes built-in make attractive seating. This project looks complicated, but the plans and step-by-step instructions put your basic skills to work.

BASIC WORKBENCH

Materials List
Top (2), 24″ × 60″ × ³/₄″ plywood
Shelf (1), 19¹/₂″ × 48″ × ³/₄″ plywood
Top side braces (2), 2″ × 4″ × 21″ lumber
Top front and back braces (2), 2″ × 4″ × 45″ lumber
Bottom side braces (2), 2″ × 4″ × 18″ lumber
Bottom front and back braces (2), 2″ × 4″ × 45″ lumber
Sides (2), 21″ × 27″ × ¹/₄″ plywood
Back (1), 27″ × 48″ × ¹/₄″ plywood
No. 14 × 3″ flat head wood screws
6d nails
White glue
Wood putty (optional)

Tools
Handsaw or circular saw
Drill
Screwdriver
Hammer
Power sander or sandpaper
Nailset (optional)

Length of Time Required
Afternoon

Safety Precautions
Safety goggles

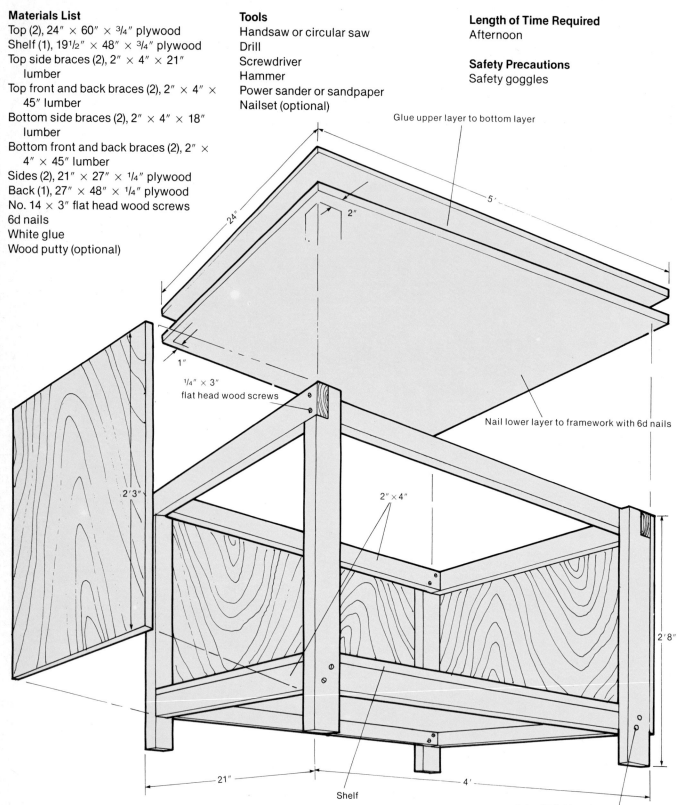

Glue upper layer to bottom layer

24″

5′

2″

1″

¹/₄″ × 3″ flat head wood screws

Nail lower layer to framework with 6d nails

2′3″

2″ × 4″

2′8″

21″

4′

Shelf

¹/₄″ × 3″ flat head wood screws

Exploded View

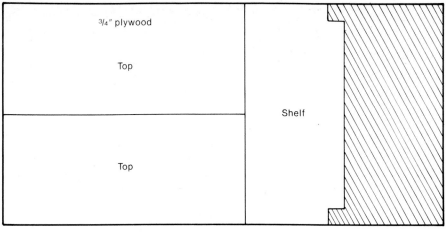

Panel Diagram

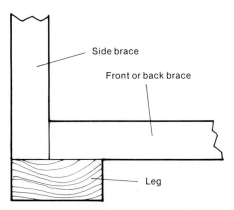

Detail A: Bottom Brace Corner

A sturdy workbench with ample surface area is an indispensable furnishing in the woodworking shop. The following project is a basic workbench, one suitable for the beginning woodworker. It is constructed of plywood (³/₄-inch for the top and shelf, ¹/₄-inch for the sides and back) and 2 × 4s.

To be most useful, your workbench should be the right height for you—the bench top should be the height of your hip joint. The plans given here are for a bench 33¹/₂ inches high. Adjust the plans to match your height by increasing or decreasing the length of the leg and, consequently, the vertical dimensions of the back and sides.

Once you have completed the bench and started to use it, you will discover whether you prefer to work at one side of it, with the bench pushed up against a wall, or around all four sides, with the bench standing in the center of your shop. If you place it against a wall, be sure that you have at least 2 feet of clearance at either end for large work. Lack of clearance will cut down on the usable surface area of the bench. If you place the bench in the center of your shop, make arrangements to supply it with accessible outlets, either in the ceiling or floor, so that you don't have to run extension cords. These can trip you up as you move around the bench.

STEP 1
Cutting the parts and assembling the frame ends
Mark a 4 × 8 sheet of ³/₄-inch plywood according to the dimensions given for the top and shelf (see the Panel Diagram) and cut out the pieces with a handsaw or circular saw. Notch the corners of the shelf to fit around the 2 × 4s (actual dimensions are 1¹/₂ × 3¹/₂), as shown. Mark and cut the sides and back from ¹/₄-inch plywood and cut the legs and braces to the given dimensions. Notch one end of each leg 3¹/₂ inches deep and 1¹/₂ inches in to accept the top side braces in a half lap joint. Pre-drill and countersink screw holes for the top side braces and complete the joints for the two frame ends with glue and No. 14 × 3-inch flat head wood screws. Then fasten the bottom side braces between the legs with glue and No. 14 × 3-inch flat head wood screws (see the Assembly View).

STEP 2
Completing the frame
Use glue and 6d nails to attach a side panel to one of the frame ends. Then use glue and No. 14 × 3-inch flat head wood screws to attach the front and back braces between the ends at the top and bottom (see the Exploded View and Detail A). Do not affix the other side panel until the shelf is installed.

STEP 3
Completing the bench
Install the lower shelf in the frame through the open end and glue and nail it in place using 6d nails. Center one of the top panels on the frame and fasten it to the frame with glue and 6d nails. Attach the back panel and other side panel in the same manner. Finally, glue the remaining top panel onto the one already fastened to the frame, weight it evenly with fairly heavy items, and allow it to dry overnight. Using only glue gives you a surface free of nail heads and screw holes. Of course, for extra security you can use 6d finishing nails, set and filled.

BACKYARD PLANTER BENCHES

Materials List

No. 10 × 2½" flat head wood screws
4d galvanized finishing nails
Wood preservative (pentachlorophenol,
    or some other)
Wood putty
Waterproof adhesive (acrylic, resorcinol)
Penetrating stain

Straight Bench

* 1 sheet, 4' × 8' × ⅜" EXT plywood
    siding
† 5 pieces, 2" × 4" × 8' lumber
    2 pieces, 2" × 4" × 6' lumber
    4 pieces, 2" × 2" × 8' lumber
    2 pieces, 2" × 2" × 6' lumber
    1 piece 1" × 3" × 6' lumber
    1" dowel × 2'

L-Shaped Bench

* 1 sheet, 4' × 8' × ⅜" EXT plywood
    siding
† 11 pieces, 2" × 4" × 8' lumber
    7 pieces, 2" × 2" × 8' lumber
    4 pieces, 2" × 2" × 6' lumber
    2 pieces, 1" × 3" × 6' lumber
    1" dowel × 4'

* The siding shown is APA type 303, which
    comes in a variety of textures.
† Use Douglas fir, pine, or spruce for all lumber.

Tools

Stationary power saw, circular saw, or
    handsaw
Square
Hammer
Screwdriver
Hand brace or power drill
Carpenter's level
Straightedge
Nailset

Length of Time Required

A few days

Safety Precautions

Safety goggles

This pair of backyard benches will solve your outdoor seating problems and let you show off your green thumb in the bargain. Build one or both—the construction is essentially the same. You can modify the design: put planters at both ends of the straight bench. Or design your own bench—with another corner box, it would be a simple matter to make the L-shaped bench into a U. The bench tops are of 2 × 4 and 2 × 2 lumber, with pieces spaced 1 inch apart to allow rain or lawn-sprinkling water to drain through.

The benches can be built using only hand tools, but a stationary or portable power saw is best for both accuracy and speed, especially when cutting the many miter joints in this project. In addition to appearance, the miter joints protect the raw edges of both lumber and plywood. But if you don't have access to a power saw (or if your plywood miters aren't perfect even with a power saw), all is not lost. Just cover the box corners with outside corner molding—and no one (except you) will ever know the difference.

STEP 1
Cutting the plywood

Mark the plywood sheet (or sheets) as shown in the Panel Layout, allowing ¼ inch extra at edges of pieces to be mitered. Miters are cut on the 15-inch edges of planter sides A and

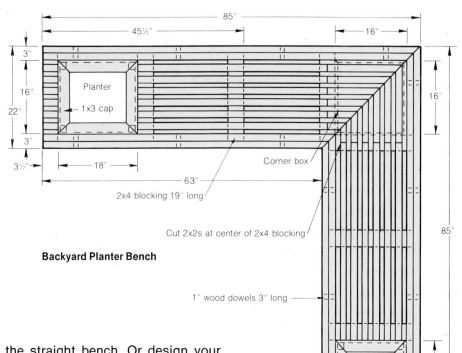

Backyard Planter Bench

B and on the 14⅝-inch edges of box sides C and D. Cut out the plywood pieces, mark identifying letters on the back sides, then trim the mitered edges at a 45° angle to the correct size.

STEP 2
Cutting and treating the lumber

This part of the project should be done well in advance to give the lumber ample time to absorb the wood preservative. Soak the base pieces in preservative for two weeks (you can buy pre-treated materials, of course) and use when dry enough to handle. Treat the interior sides and bottoms of planter boxes with wood preservative (follow label directions). But before treating, cut the lumber. Take

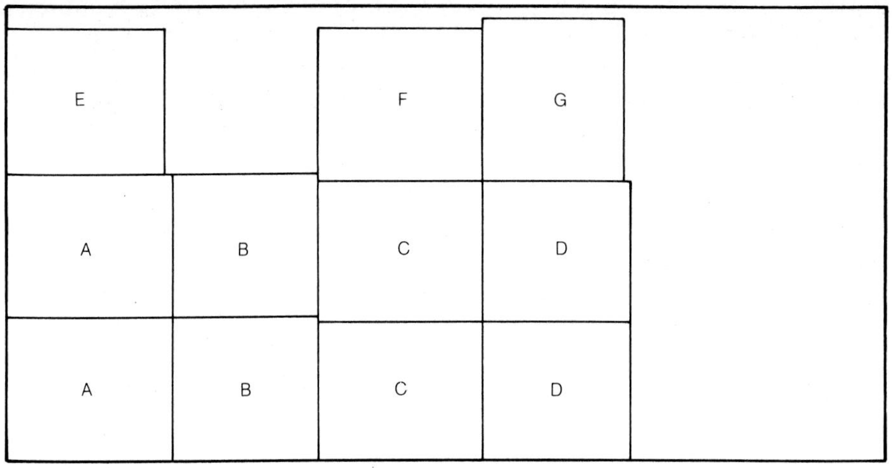

E

A   B   C   D

A   B   C   D

Straight bench

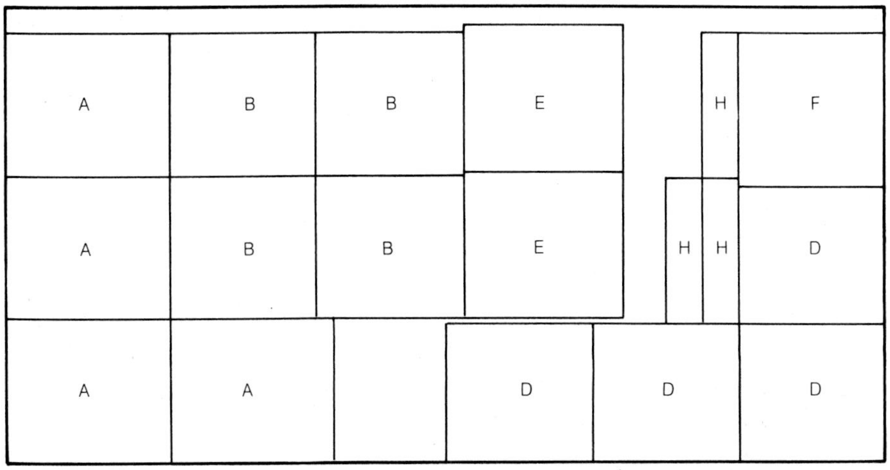

A   B   B   E   H   F

A   B   B   E   H   H   D

A   A   D   D   D

L-shaped bench

Panel Layout

⅜" x 4' x 8' EXT-
APA 303 plywood siding

A - planter side 15" x 18"
B - planter side 15" x 16"
C - box side 14⅝" x 18"

D - box side 14⅝" x 16"
E - planter bottom 15¼" x 17¼"
F - box top 16" x 18"
G - box bottom 15¼" x 17¼"
H - box top 16" x 16"
I - box bottom 4" x 15¼"

great care in laying it out to minimize waste.

For the **straight bench**, cut the following.

- from the 8-foot 2×4s—four 85-inch aprons; five 19-inch block supports
- from the 6-foot 2×4s—four 17¼-inch base pieces (miter both ends); four 15¼-inch base pieces (miter both ends)
- from the 8-foot 2×2s—one 63½-inch bench top; one 12⅝-inch planter corner; one 12¼-inch box corner
- from the 6-foot 2×2s—two 63½-inch bench tops
- from 2×2 scrap—six 3½-inch bench ends

For the **L-shaped bench**, cut the following.

- from four 8-foot 2×4s—two 85-inch outer aprons (miter one end); two 83½-inch outer aprons (miter one end)
- from four 8-foot 2×4s—two 66-inch inner aprons (miter one end); two 64½-inch inner aprons (miter one end); four 17¼-inch base pieces (miter both ends)
- from two 8-foot 2×4s—eight 19-inch block supports; two 15¼-inch base pieces (miter both ends)
- from one 8-foot 2×4—six 15¼-inch base pieces (miter both ends)
- from two 8-foot 2×2s—two 59½-inch bench tops (miter one end); four 12¼-inch box corners
- from four 8-foot 2×2s—two 57-inch bench tops (miter one end); two 54½-inch bench tops (miter one end); eight 12⅝-inch planter corners

- from one 8-foot 2×2—two 47-inch bench tops (miter one end)
- from the 6-foot 2×2s—two 52-inch bench tops (miter one end); two 49½-inch bench tops (miter one end)
- from 2×2 scrap—twelve 3½-inch bench ends

STEP 3
Assembling the planters and end boxes

The bases for the planter and end boxes both measure 15¼ × 17¼ inches, while that for the corner box in the L-shaped bench is 15¼ inches square. Assemble the bases with glue and two nails driven in each direction at each corner, making certain that all corners are square.

Planter sides (A and B) are 15 inches high; the 2×2 corner posts are 12⅝ inches long. End and corner box sides (C, D) and corner posts are 14⅝ and 12¼ inches, respectively. To assemble, line up the mitered edge of a side along one edge of a corner post. Spread adhesive along the 2×2 post and nail on the plywood side, with its top flush with the top of the post, spacing the nails about every 3 inches. Repeat gluing and nailing at all corners, making

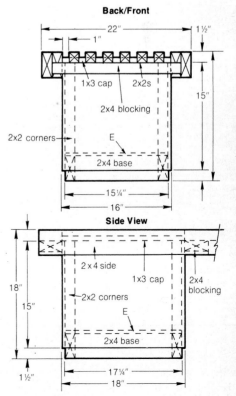

Back/Front

22"   1½"
1"
1x3 cap   2x2s
2x4 blocking
2x2 corners   E
2x4 base
15"
15¼"
16"

Side View

2 x 4 side
1x3 cap   2x4 blocking
2x2 corners
E
2x4 base
18"
15"
1½"
17¼"
18"

sure that the unit is square. When the box shell is complete, insert the bottom piece, nailing to the corner posts. (*Note:* In the corner storage unit, nail the two outside bottom pieces to the posts and nail the middle piece to the base.) Set the box on the assembled base, nailing through the sides into the 2×4s.

STEP 4
Starting bench assembly
Aprons for the bench tops are made up of doubled 2×4s. The inner 2×4s are notched 2 × 3½ inches to receive the support blocks (see the drawings). Drill pilot holes through the blocks to attach the 2×2 bench top pieces with screws. Use one

screw at each intersection and two screws at the short end pieces alongside the planter. For the storage box, cut through the 2×2s over the 2×4 support just outside the box. Then glue and nail the 2×2s to the box top. Add the inner apron pieces, driving screws through the support blocks into them. Drill 1-inch holes for dowels through both outer and inner apron pieces at locations indicated on the drawings. Glue the apron pieces together, then drive 3-inch lengths of glue-coated dowel into the holes, flush with the outer surface of the apron.

STEP 5
Completing bench assembly
If the bench is to be located on a patio, set the boxes in place and add the bench top. If it is placed in a yard, set the boxes in place and roughly level and align them using a straightedge and a carpenter's level. Place the top on the boxes and check with the level. If necessary, remove a little soil from beneath the base or shim with stones or brick chips to make the boxes level and solid. Nail through the inner edges of the

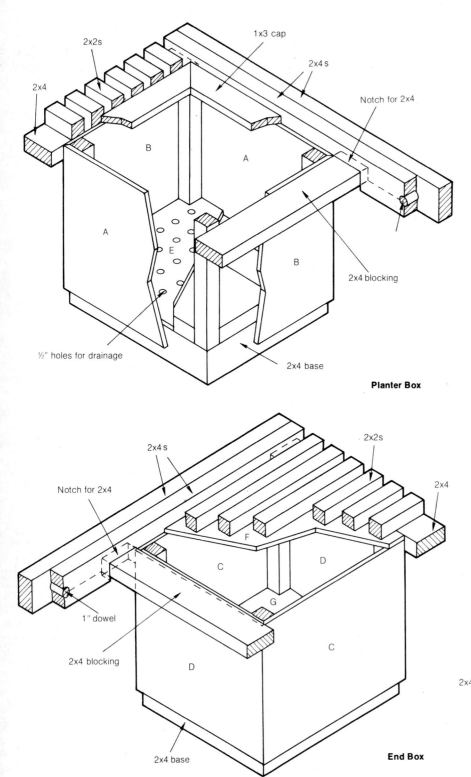

Planter Box

End Box

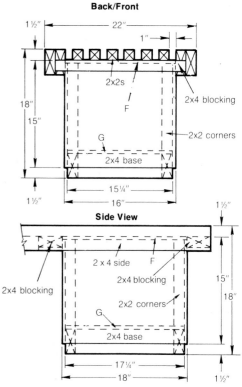

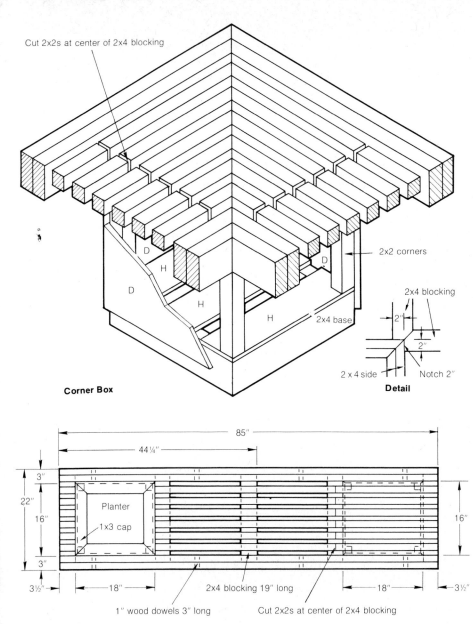

Cut 2x2s at center of 2x4 blocking

2x2 corners

D

H

D

H

H

2x4 base

Corner Box

2x4 blocking

2"

2"

2 x 4 side

Notch 2"

Detail

85"

44¼"

3"

22"

16"

3"

Planter

1x3 cap

16"

3½"

18"

2x4 blocking 19" long

18"

3½"

1" wood dowels 3" long

Cut 2x2s at center of 2x4 blocking

planter box into the surrounding 2×4s. If a storage box is used at the other end, nail through its inside into the 2×4 supports and aprons. (If a solid end box is used rather than a storage unit, the top can simply rest on it—no need to fasten it.)

STEP 6
Finishing the bench

Cut 1×3 cap pieces (two 18-inch, two 16-inch) for the planter box and miter the corners. Run a bead of glue around the top edges of the planter, then attach the cap pieces, gluing the mitered edges and nailing through them into the corner posts; toe nail into the 2×4 blocking and aprons at 3-inch intervals.

Countersink all exposed nails with a nailset and fill the holes with wood putty. Sand smooth. Apply stain, following manufacturer's directions. If you wish, you may apply a protective coating, such as an exterior-type varnish, to the bench top. Clear finishes on plywood exposed outdoors are unsatisfactory and should be avoided.

LUMBER CHAIR

Materials List

* 3 pieces, 2″ × 8″ × 7′ lumber
  1 piece, 2″ × 6″ × 7′ lumber
  1 piece, 2″ × 6″ × 5′ lumber
  50 No. 10 × 2½″ flat head wood
    screws
† 30 ½″ birch furniture buttons
‡ White glue
  Clear finish
  Seat cushion (1), 21″ × 21″ × 3½″
  Back cushion (1), 18″ × 21″ × 3½″

* For all lumber, use Douglas fir, pine, or spruce; Select or Knotty, as you prefer. If you choose the latter, make sure the knots are tight.
† You can substitute ½″ dowel to cover the screw heads.
‡ Use waterproof resorcinol glue if the chair is to be used outdoors or in extremely damp conditions.

Tools

Stationary power saw, circular saw, or
  handsaw
Saber saw or keyhole saw
Scissors
Hand brace or power drill
Screwdriver
Square
Forming tool
Sanding block or power sander

Length of Time Required
Overnight

Safety Precautions
Safety goggles

Lumber Chair

Building furniture need not always involve complicated cutting and forming with expensive power tools. The basic materials for this chair come right off the rack at your lumberyard—all standard sizes that need only to be cut to length. If you choose to, arrange for your lumber dealer to make the cuts for you on his shop saw. Otherwise, you can easily make them yourself with either a handsaw or a power saw. Just make sure that all cuts are perfectly square. This project and the matching table that follows offer a good opportunity to try out basic skills.

STEP 1
Cutting the parts

Mark and cut the lumber to the following dimensions: for the legs—four pieces, each 2 × 8 × 28 inches, and two pieces, each 2 × 6 × 28 inches; for the sides—four pieces, each 2 × 8 × 21 inches; for the back—two pieces, each 2 × 8 × 21 inches; and for the seat—four pieces, each 2 × 6 × 21 inches.

Use a compass and pencil to draw a circle with a 2-inch radius on a piece of cardboard. (If you don't have

a compass handy, use a jar, a vase, a funnel, a mug—anything round that measures 4 inches across.) Cut out the circle with a scissors, then use the curve of this pattern to mark the top front edges of all four 2 × 8 legs and two of the sides. Cut the curves with a saber saw or a keyhole saw.

STEP 2
Pre-drilling the parts

The chair is assembled with glue and screws. Three equally spaced screws are used to attach each 2 × 8

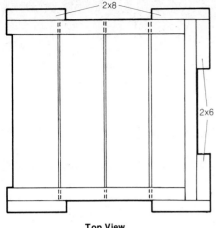

Top View

Drill screw holes ¼ deep (note tape on drill bit to mark required depth).

Side View

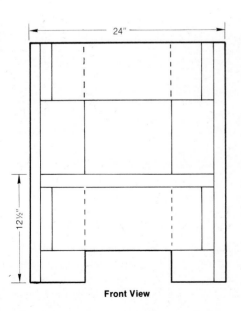

Front View

STEP 3
Assembling the chair

Begin assembly by joining the back corner leg sections, gluing and screwing the 2×8s to the 2×6s. Use a damp cloth to wipe away any excess glue that is squeezed out of the joint. Follow this practice throughout the assembly.

Glue and screw the front legs to the top side pieces, carefully checking the alignments with a square. Attach the upper back piece to the top side pieces. Don't forget to wipe away excess glue.

Fasten the back bottom to the two lower side pieces. Starting flush with the back edge of the back bottom, glue and screw the four seat pieces to this assembly. Position the assembled seat unit inside the front legs, 12½ inches from the bottom; square it up and fasten. Add the back corner legs, again carefully checking to make sure that everything is kept in square.

STEP 4
Forming and sanding

Use a forming tool and a power sander or sanding block to smooth any irregularities where the front leg and side curves are joined. Sand to ease all cut or sharp edges on the

side leg to the 2×6 back legs; all other joints are made with two screws. Except for the seat and the back bottom, screws at these joints should be offset and aligned, for example, at front and back legs, as well as on both sides (see the Side View for locations). Carefully mark the locations for all screws. With a ½-inch bit in a hand brace or power drill, drill holes, ¼ inch deep, for the screws. (Make a depth gauge by wrapping the bit with tape ¼ inch above the cutting edge.) Then use a 3/16-inch bit to drill a pilot hole for the screw shank through the middle of each ½-inch hole.

Glue and screw the seat boards to the frame, starting flush with the back.

Sand the chair thoroughly with fine abrasive paper before finishing. Note that the sander should be moved in the direction of the grain.

Use a forming tool and a sander to smooth curves where legs and sides meet.

Glue wood buttons into the screw holes, taking care not to use too much glue and wiping away any excess. The chair is ready to finish.

chair, to remove any excess glue that you neglected to wipe away during assembly, and to clean away any stains or fingerprint smudges that may have been left on the wood. Glue wood buttons into all exposed screw holes (except those in the seat—you can leave these exposed).

STEP 5
Finishing the chair

Apply a clear finish to the chair, following the manufacturer's instructions. You could, of course, stain the chair; but if you were careful in selecting the lumber, why hide it?

Either make cushions to fit from block foam and fabric (see the Materials List for overall measurements) or have an upholstery shop make them to your specifications. When the finish is dry, pop the cushions in place, sit back, and enjoy your new chair!

LUMBER COFFEE TABLE

Materials List
* 2 pieces, 2″ × 8″ × 6′ lumber
  2 pieces, 2″ × 4″ × 6′ lumber
  1 piece, 2″ × 4″ × 7′ lumber
  32 No. 10 × 2¹/₂″ flat head wood screws
† ¹/₂″ dowel × 10″
‡ White glue
  Clear finish

* For all lumber, use Douglas fir, pine, or spruce; Select or Knotty, as you prefer. If you choose the latter, make sure that the knots are tight.
† You can substitute wood buttons, as shown in the Lumber Chair.
‡ Use waterproof resorcinol glue if the coffee table is to be used outdoors or in extremely damp settings.

Tools
Stationary power saw, circular saw, or handsaw
Hand brace or power drill
Screwdriver
Miter box and back saw
Sanding block or power sander

Length of Time Required
Overnight

Safety Precautions
Safety goggles

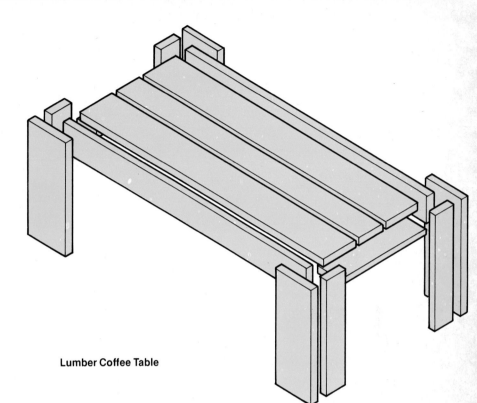

Lumber Coffee Table

For drinks, snacks, magazines, and any other amenities of relaxation, this coffee table is the perfect complement to the preceding lumber chair. Or you might just build it as an individual piece—it will hold its own with any number of furniture styles. As is the chair, the table is put together with off-the-rack materials.

STEP 1
Cutting the parts
You can have the lumberyard cut all the pieces to length, or you can make the cuts yourself with either a handsaw or a power saw. Make sure that all the cuts are square.

Mark and cut the lumber to the following dimensions: for the legs—four pieces, each 2 × 8 × 14³/₄ inches, and four pieces, each 2 × 4 × 14³/₄ inches; for the top—two pieces, each 2 × 8 × 41 inches, and one piece, 2 × 4 × 41 inches; for the apron—two pieces, each 2 × 4 × 41 inches; and for the end supports—two pieces, each 2 × 4 × 18 inches.

STEP 2
Assembling the leg corners
The coffee table is assembled with glue and screws. Mark screw locations on the legs as follows (check the drawings for locations): on the 2 × 8 legs, screws into the 2 × 4 legs are located ³/₄ inch in from the edge, 2 inches from both top and bottom; a third screw, 1 inch in from the opposite edge and 1¹/₂ inches from the top, goes into the apron. Mark the 2 × 4 legs 1 inch in from the edge, 2 inches from the top, for a screw into the apron. With a ¹/₂-inch bit in a power drill or a hand brace, drill holes, ¹/₄ inch deep, at the screw marks (wrap tape around the bit to serve as a depth gauge). Drill ³/₁₆-inch holes through the pieces in the middle of the ¹/₂-inch holes. Apply a bead of glue to the 2 × 4 leg pieces and assemble the leg corners by driving screws through the 2 × 8s into the 2 × 4s. Wipe away excess glue from the joints with a damp cloth.

STEP 3
Assembling the top
Square the top pieces—2 × 8s flanking the 2 × 4—with the end supports. Drill pilot holes through the supports into the undersides of the top pieces (two into each piece at each end) and, with glue and screws, fasten securely. Since these screws will be concealed beneath the table, it is only necessary to drive them snug; you need not countersink them.

STEP 4
Final assembly
Drill ¹/₂-inch holes, ¹/₄ inch deep, in the apron pieces, ³/₄ inch from the top edge and 15 inches from each end. Then drill ³/₁₆-inch holes through the middle of the ¹/₂-inch holes. With glue and screws, fasten

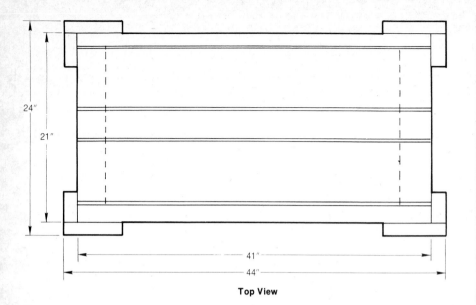

24″
21″
41″
44″

Top View

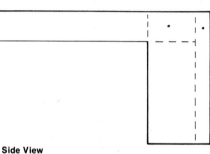

14¾″

Side View

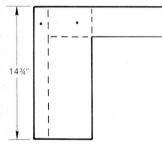

Square up leg sections at each corner, then attach with glue and screws.

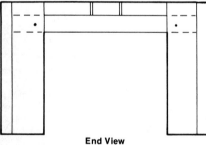

End View

the aprons to each side of the top. Wipe away excess glue.

Add leg sections at each corner, carefully squaring the unit before driving in the screws.

STEP 5
Finishing the table

In a miter box, cut dowels to approximately ½-inch lengths with a back saw. Coat them with glue and tap into all visible screw holes. Let glue dry. Cut the dowels flush with the surface, using a sharp chisel or a fine-tooth saw; sand smooth. Sand any rough spots, stains, or finger smudges. Apply a clear finish, following manufacturer's directions.

EARLY AMERICAN-STYLE WALL SHELF

Materials List

1 sheet, 4' × 4' × ½" A-A or A-B plywood

1 piece, ½" × ½" × 8' hardwood or pine strip

4d finishing nails

1" brads

Glue (urea-resin type)

Wood putty

* Stain

Clear finish

* Maple is popular for this type of project, but choose what best suits your taste.

Tools

Stationary power saw, circular saw, or handsaw

Band saw, coping saw, saber saw, or jig saw

Circular saw with dado blade (optional)

Hammer

Nailset

Sanding block or power sander

Length of Time Required

Overnight

Safety Precautions

Safety goggles

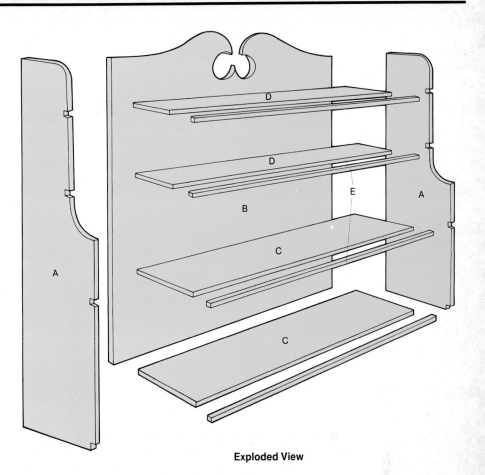

Exploded View

Display your favorite knick-knacks, books, small plants, or anything else that suits your fancy on this handsome Early American-style shelf unit that hangs on the wall. You can build it with ordinary hand tools, although a power saw and an electric drill will make the job easier.

STEP 1
Cutting the parts

Lay out plywood pieces A through D as shown on the Panel Layout. Face grain of the plywood should run the long way of A and B. Cut shelves C and D first, then separate sides A from back B. Lay out the scroll on the top of back B, following the grid in the Cutting Diagram. The best tool for cutting the scroll is a band saw,

but a coping saw will do it almost as well; a saber saw or jig saw may also be used.

STEP 2
Completing the sides

Cut apart two sides A and lay out one side according to the drawing, using a compass to mark the 2¼-inch inside and outside radii. Tack the two sides together, good sides out, and cut to shape with a saber saw. (A handsaw may also be used, with a keyhole saw or coping saw to cut the curves.) The ¼-inch-deep × ½-inch-wide notches at each shelf may be cut with a ½-inch dado blade in a circular saw; or you can mark them, separate the two sides, and cut them with a coping saw. Fill in any rough

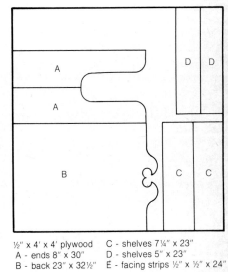

½" x 4' x 4' plywood

A - ends 8" x 30"
B - back 23" x 32½"
C - shelves 7¼" x 23"
D - shelves 5" x 23"
E - facing strips ½" x ½" x 24"

Panel Layout

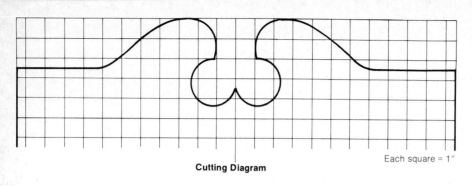

Cutting Diagram

Each square = 1"

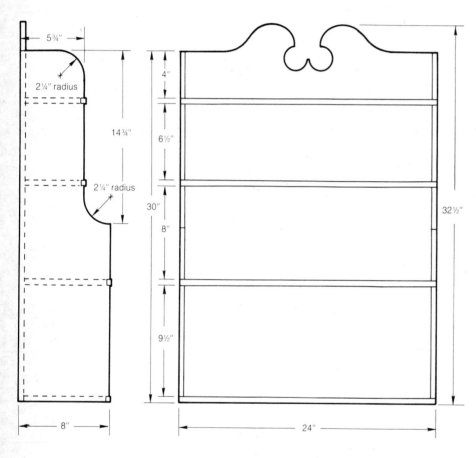

through the back. Install remaining shelves, aligning them with the marks on sides and back. Wipe away excess glue after each shelf is nailed in place.

STEP 4
Adding facing strips
The facing strips (E) can be cut with a back saw in a miter box, but any careful, square sawing will do. Glue strips E into the notches and along the front edges of each shelf and fasten to the front edges of the shelves with 1-inch brads at each end and in the middle.

STEP 5
Finishing the shelf
Use a nailset to drive all nail and brad heads beneath the surface. Fill all the holes with wood putty and sand smooth. Complete any other touch-up sanding that may be required, working in the direction of the grain.

Apply stain according to manufacturer's directions. Then apply a clear finish to protect the surface of the wood.

STEP 6
Mounting the shelf
The shelf unit can be mounted on the wall by driving 2-inch wood screws into wall studs through holes drilled just above the bottom shelf and the top shelf—the 24-inch width should make it easy to catch two studs on 16-inch centers. Round head screws will be acceptable, since they will probably be concealed behind whatever you display on the shelves. If you want to go that extra step, you can countersink flat head screws (but not too deeply, since you might weaken the holding power of the back) and fill the holes with wood putty. Then touch up with stain to match the unit.

If your shelf placement makes it difficult to find studs in the right position for fastening, use hollow-wall fasteners (see *Fasteners and Adhesives* for types and applications).

edges on the top, front, and bottom of each side with wood putty. Sand smooth.

STEP 3
Assembling shelf unit
When assembling, always make sure that good sides of the plywood face outward. Mark locations of shelves on sides and back. Run a bead of glue down one edge of back B and at-tach one side A with three 4d finishing nails, holding the two pieces flush at top, back, and bottom. Wipe away any excess glue squeezed out of the joint. Repeat the procedure to attach the other side to the back. Apply glue to the side and back edges of bottom shelf C and set in place flush with the bottom of assembly A–B–A. Secure it with two 4d nails driven through the sides and two

HINGED BOX

Materials List
1 piece, $\frac{1}{2}'' \times 8'' \times 30''$ stock lumber
    (select for grain pattern)
1″ brads
White glue
Wood putty
Stain (optional)
Lacquer
Hardware (shown, Stanley Classic
    Brassware):
    CD5347 #1 top corners (4)
    CD5348 #1 bottom corners (4)
    CD5302 $\frac{3}{4}''$ hinges (2)
    CD5328 hasp (1)
Velvet remnant, 9″ × 18″ (optional)

Tools
Combination or try square
Fine-tooth power saw, handsaw, or
    crosscut saw
Hammer
Nailset
Screwdriver
Sandpaper

Length of Time Required
Afternoon

Safety Precautions
Safety goggles

This small hinged box makes a handsome dresser-top repository for jewelry, cuff links, and countless other small items. It looks like (and is) a finely crafted piece, but there's a bit of magic in its construction—turning a solid cube into a perfectly fitting box and cover—that makes the job a lot easier than the finished product looks. The hardware used in this project was manufactured by Stanley, and the pieces are listed here with their order numbers; but any comparable hardware will work just as well.

STEP 1
Cutting the pieces
Use a combination or try square to measure and mark the $\frac{1}{2} \times 8$ into one 17-inch and one 13-inch segment. Make the cut with a fine-tooth power

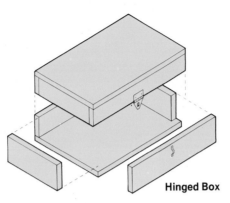

Hinged Box

or crosscut saw. Rip the longer piece to a width of $5\frac{1}{2}$ inches, then cut it into two 8-inch lengths (for the box top and bottom). Rip the 13-inch length to two $2\frac{7}{8}$-inch widths, then cut these to lengths of 8 inches (for the box front and back) and $4\frac{1}{2}$ inches (for the ends).

STEP 2
Assembling the cube
Glue the ends between the front and back, then fasten with brads no more than $\frac{1}{2}$ inch from the outside edge (see the End View). With glue and brads, attach the top and bottom to the front, back, and ends, forming a closed cube. Set all brads and fill the holes and any voids in the joints with wood putty. Sand the cube and smooth all sharp edges.

STEP 3
Cutting apart the cube
Draw a line $1\frac{1}{4}$ inches from the top all around the front, ends, and back of the cube. With a fine-tooth handsaw, cut just below the line, first through one end, then the front, then the other end. Make the final cut through the back to separate the cover from the box. You can also make the cuts on a table saw, setting the ripping fence for $1\frac{1}{4}$ inches and the blade for a cut $\frac{5}{8}$ inch deep. As above, cut end, front, and end; then insert a $\frac{1}{8}$-inch filler strip in the front cut to prevent binding and make the final cut in the back.

STEP 4
Completing the box
Sand the cut edges smooth. Apply stain if desired, then lacquer to give the chest a lustrous finish. When dry, attach hinges on the back, 1 inch from each end. Attach the hasp and the top and bottom corners, using the screws supplied with the sets.

If you wish, you can further protect your treasures by lining the box with velvet. Cut two pieces, one approximately $9 \times 11\frac{1}{2}$ inches for the box and one $9 \times 6\frac{1}{2}$ inches for the cover. Fit them in place, mitering the corners and cutting away excess material. After fitting, glue the velvet in place. When the glue has dried, trim away the excess with a sharp single-edge razor blade.

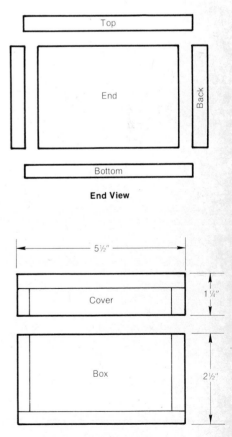

End View

Box (after cutting apart)

MODULAR CUBES

Materials List

1 sheet, ½" lumber core birch plywood, good both sides
Brads
No. 4 × ³⁄₈", No. 4 × ⁵⁄₈" flat head wood screws
White glue
Butt hinges with mounting screws (3), 1½" × 1"
Magnetic catches with mounting screws (3)
Brass knobs (3), ½" diameter
Aluminum screw post fasteners (8), 1" long
Stain

Tools

Table saw, radial arm saw, hand saw
Drill
Hammer
Screwdriver
Sander or sandpaper

Length of Time Required

Afternoon

Safety Precautions

Safety goggles

These cubes are little more than boxes, although they are elegant boxes. Each is made with lumber core birch plywood and assembled with simple hardware. The more you build, the more storage space you have. Once you have built a few, you should be able to produce them very quickly.

As mentioned, birch plywood is used. Since all sides of the box are finished, you will have to purchase plywood with two good veneer faces. To save money, you could use fir plywood and finish the boxes with paint.

These instructions and the materials listed will produce three cubes with doors.

STEP 1
Cutting the parts

Cut the sheet of birch plywood into 15¾ × 15¾-inch squares; one sheet will yield exactly eighteen pieces.

Pick three of the pieces for backs, and cut them to a finished size of 15¼ inches square. Select three additional pieces for the doors, and trim them to 14¹¹⁄₁₆ inches square.

STEP 2
Trimming and rabbetting the sides, tops, and bottoms

Trim six of the 15¾-inch square pieces to 15¼ × 15¾ inches. Cut a ¼-inch-deep by ½-inch-wide rabbet along one of the 15¼-inch sides of each of the six pieces.

Take the remaining six 15¾ inch square pieces and rabbet each piece ¼ inch by ½ inch on three sides; these parts form the top/bottom portion of the three cubes.

STEP 3
Assembling the cubes

Using white glue and brads, assemble one side to a top or bottom piece

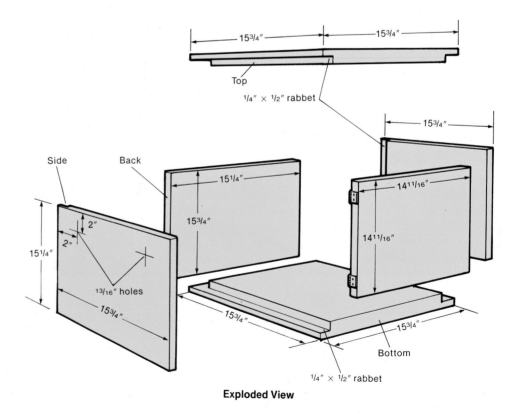

Exploded View

and to the back. Then assemble the remaining side and top/bottom for a completed open cube. Set the nailheads and fill the holes with wood putty. Repeat these steps for the remaining two cubes.

STEP 4
Drilling joining holes
Mark the three cubes for the joining holes, as indicated in the Exploded View. To ensure alignment of the holes from cube to cube, lay out the hole pattern on a master drawing and transfer these hole positions to the units using a sharp instrument. Drill the $^{13}/_{64}$-inch diameter holes required.

STEP 5
Attaching the hardware
Lay out all hardware mounting positions as shown in the Exploded View. The magnetic catch should be centered, but the exact fastener positions will depend on the particular catch you buy.

Use No. 4 × ½-inch flat head wood screws to mount the hinges to the doors, and No. 4 × ³/₈-inch flat head wood screws to join the hinges

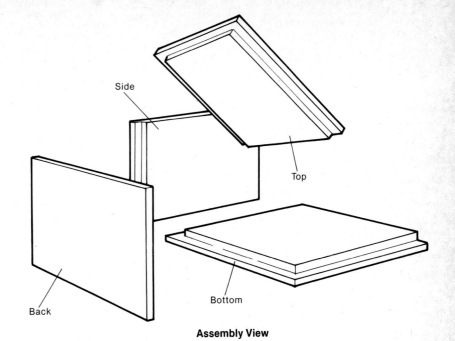

Assembly View

to the cube. Mount, then remove the hinges before finishing the wood.

STEP 6
Finishing the cubes
The surface of cabinet-grade birch plywood is excellent, so your stack cubes will require only light sanding with extra fine sandpaper before ap-

plying your finish, preferably a wood stain that will bring out the pattern and grain of the birch.

Reattach the hinges and install magnetic catches. Join the cubes with aluminum screw post fasteners; these are available at office supply stores.

BABY'S CHANGING TABLE

Materials List

Uprights (4), 1½″ × 1½″ × 32″, any
  stock
Sides (2), ¾″ × 16″ × 31″,
  hardwood-faced plywood
Back (1), ¾″ × 31″ × 37″,
  hardwood-faced plywood
Shelves (4), ¾″ × 16″ × 37″,
  hardwood-faced plywood
Facers (4), ¾″ × 2″ × 37″, any solid
  stock or hardwood-faced plywood
Wood tape, 8′ required
1½″ spiral dowels (24)
8d finishing nails
Casters
Glue
Wood putty
Varnish or lacquer

Tools

Table saw or radial arm saw
Doweling jig
Drill
Sander or sandpaper
Hammer
Clamps

Length of Time Required

A few days

Safety Precautions

Safety goggles, filter mask

A changing table or dressing stand not only provides the right working height for changing and dressing a baby, but also the convenience of storage space for such things as powders, diapers, clothes, and other necessary items. This table is designed to accommodate the standard size 2 × 16 × 36-inch pad available in stores that sell baby furniture. Casters on the table make it easy to move.

If the changing table is constructed of a good hardwood-faced plywood or even solid stock doweled together, it can do double duty through the years: as your baby grows up, the table will change to a bookcase or set of storage shelves for the youngster's collections and other valuables. The changing table is moderately easy to build, with doweling the sides and back the only challenging operation.

STEP 1
Cutting the uprights, sides, back, and shelves

Cut the uprights from 1½-inch stock to the proper dimensions. Then use a table saw or a radial arm saw to cut a 45° chamfer around the top of each upright, as shown in the Full View. (Or cut the chamfer using a small hand plane, if you prefer.) Bore holes in the center of the bottom of the legs for the caster pins according to the dimension of the shank. Cut the back, sides, and shelves to size. Sand all the parts smooth.

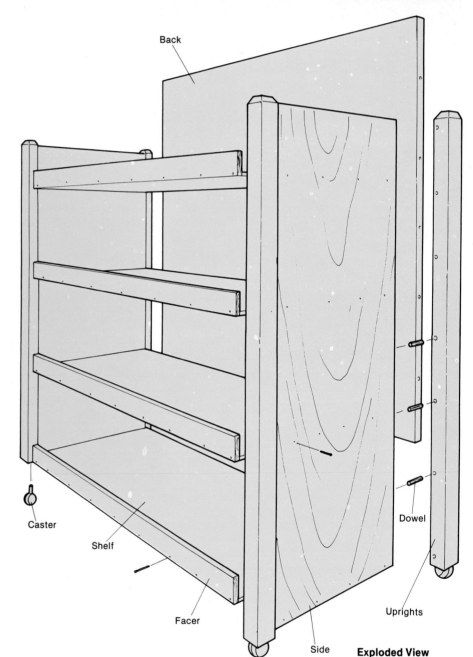

Back

Caster

Shelf

Facer

Side

Dowel

Uprights

Exploded View

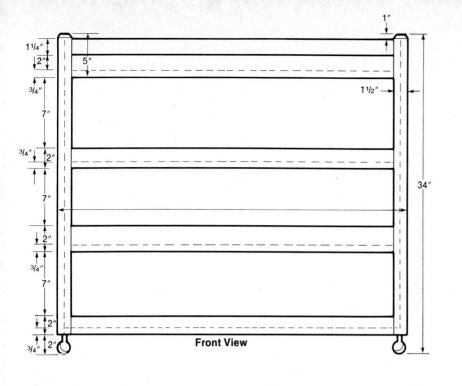

Front View

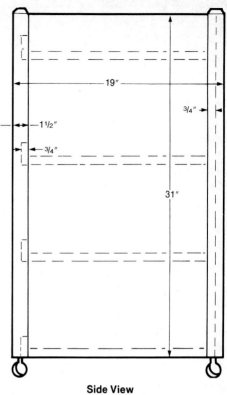

Side View

STEP 2
Making the shell

The sides and back are fastened to the uprights with glue and dowels. (You can attach the sides and back with countersunk 2-inch wood screws and glue, but this does not provide the same sturdiness as doweling.) Lay one side next to an upright in the position shown in the Side View, and carefully mark four straight lines perpendicular to the edge of the upright across the upright and side piece as a guide for the doweling jig. Mark four locations for dowels about 6 inches apart, starting 6 inches from the bottom. When the pieces are marked, use the doweling jig to bore ⅜-inch holes 1 inch deep in the center of the edge of the side and on the center line of the uprights. Repeat the operation for the other upright and the other side of

the side piece. Glue the holes and insert 1½-inch pieces of dowel into the side piece; glue the edge of the side piece; then drive an upright onto the dowels. Repeat with the other upright on the other edge of the side; then clamp the assembly together and let it dry overnight. Assemble the other side in the same manner. When the sides are set, mark, drill, and dowel the back to the back uprights, following the same procedure.

STEP 3
Installing the shelves

When the back joints are set, mark the location of the shelves inside the shell according to the dimensions on the Front View. Fasten the shelves in place to the back and sides using No. 8 finishing nails. Set the nail heads below the surface of the wood and fill the holes with wood putty.

STEP 4
Facing the shelves

Cut the shelf-facer strips to the proper size and fasten them to the front edges of the shelves with No. 8 finishing nails set below the surface and puttied.

STEP 5
Finishing

Cover the top edges of the back and sides with wood finishing tape to match the wood. This kind of tape, with peel-off backing on its adhesive side, is available at lumberyards. Sand the entire project as smoothly as possible (remember that it's for a baby) and finish with several coats of varnish or lacquer. Fit the casters on the legs.

SANDBOX

Materials List
Sides (4), 1″ × 8″ × 4′, redwood
Bottom boards (7), 1″ × 8″ × 48″, redwood
Cleats (3), 1″ × 4″ × 48″, redwood
Seat supports (12), 1″ × 8″ × 8¾″, redwood
Seats (4), 1″ × 8″ × 62″, redwood
No. 8 × 1¼″ brass round head wood screws

Tools
Table saw, handsaw, or portable circular saw
Screwdriver
Hand sander

Length of Time Required
Afternoon

Safety Precautions
Safety goggles, gloves
Filter mask

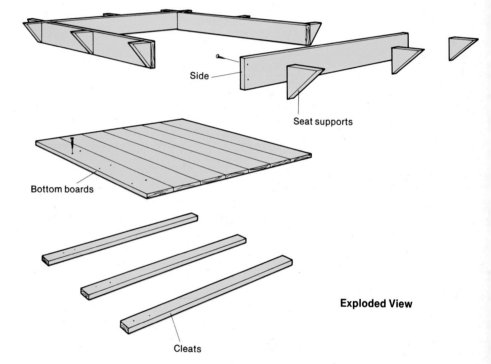

Seats

Side

Seat supports

Bottom boards

Cleats

Exploded View

A sandbox is one of the classic backyard projects, and one of the best ways to keep young children happy at play for hours. You can bring a piece of the beach into your backyard in just an afternoon. It is simply a large wooden box with a seat running around the outside; the bottom is reinforced with cleats that sit on the ground. The entire project is made with clear, heart-grade redwood 1 × 8s. The project is simple to make, but you will need help moving the finished project. When you are ready to fill the box with sand, call around to local lumberyards to find the cleanest grade of sand you can.

STEP 1
Building the box
Cut the side pieces to the given dimensions. Fasten the sides together with No. 8 × 1¼-inch brass wood screws, butting sides to create a 4-foot-square box. Cut the bottom pieces to the given dimensions and fasten them across the bottom of the box with the same-sized screws, making sure the box remains square as you work. Finally, cut the cleats to the given dimensions and fasten them with the same-sized screws across the bottom boards to keep the boards from warping and cracking. The cleats also add strength to

the whole box and keep it from twisting out of shape.

STEP 2
Adding the seats
Cut the seat supports to the dimensions given in Detail A and fasten them in place at each corner and the center of the sides with two No. 8 × 1¼-inch countersunk brass wood screws driven from the inside of the box. Be sure that the heads are well below the surface of the box and that no burrs are sticking up to snag a child's hand. For extra safety, you can fill the screw holes with plugs or wood putty. Cut the seatboards to

the given dimensions and miter the corners at a 45° angle. Fasten the seats to the edge of the box and to the supports with the same-sized countersunk wood screws driven down through the seat into the edge of the box and into the supports (two screws for each support).

STEP 3
Finishing the box

Use a hand sander to round all the edges on the seats, smooth the seats themselves, and sand the inside of the box. Place the completed sandbox on level ground, fill with clean sand, and watch the kids climb in and have a ball.

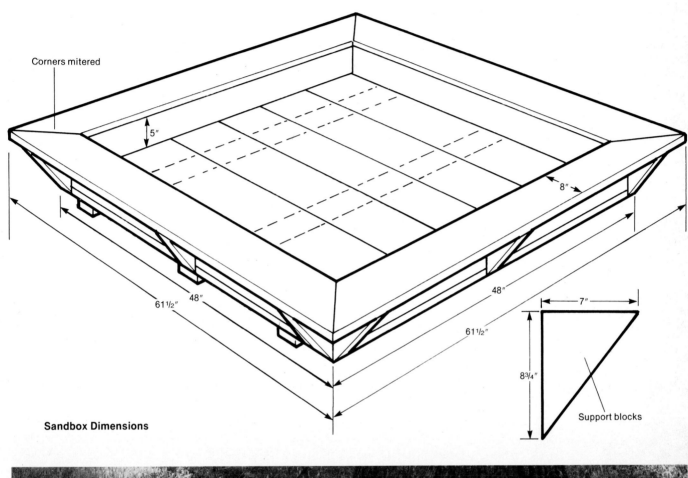

Sandbox Dimensions

SHAKER-STYLE TRESTLE TABLE

Materials List

1 sheet, 4′ × 8′ × 3/4″ A-B plywood
1 piece, 2″ × 4″ × 3′ pine or fir
2 pieces, 1″ × 3″ × 6′ pine or fir
2 pieces, 1/2″ × 3/4″ × 6′ trim
2 pieces, 1/2″ × 3/4″ × 4′ trim
1 piece, 1″ × 1″ × 24″ pine or fir
* 16″ continuous hinge, with screws
No. 10 × 2″ flat head wood screws
4 hardwood plugs (or 6″ of 3/8″ dowel)
1″ brads
Urea resin glue

Wood putty
Finish, as desired
* Four pairs of 1″ butt hinges may be
  substituted.

Tools

Handsaw or power saw
Saber saw or keyhole saw
Hacksaw
Drill
Plane or forming tool
Hammer
Screwdriver

Chisel
Power sander or sandpaper

Length of Time Required

A few days

Safety Precautions

Safety goggles

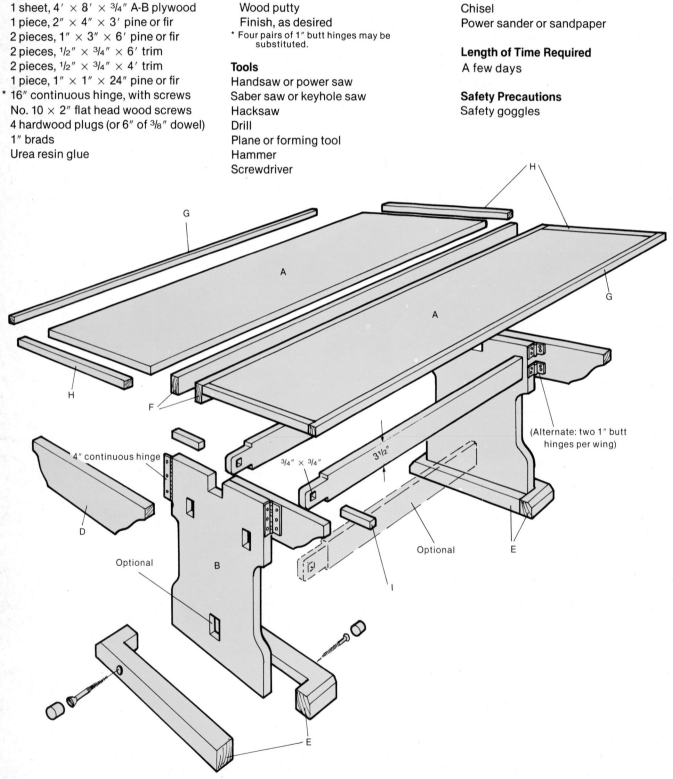

4″ continuous hinge

(Alternate: two 1″ butt
hinges per wing)

3/4″ × 3/4″

3 1/2″

Optional

Optional

Exploded View

Shaker furniture is prized for its simple, graceful lines, and this trestle table is certainly in that tradition. Its spacious surface (39½ × 66¾ inches) will serve six to eight diners or game players, or it can be used for sewing, hobbies, crafts, or whatever. It's flexible too. The two top sections can be stacked to create a refectory table. One other feature would have appealed to the practical Shakers (who believed in saving space wherever they could): simply by removing four to six pegs the table can quickly be disassembled into six or seven pieces for storage. An optional bottom stretcher is the seventh piece. It adds even greater stability to this already strong piece, but it may present an obstacle to feet when the table is used in the refectory mode.

STEP 1
Cutting the main parts
Carefully measure and mark the plywood as shown in the Panel Layout.

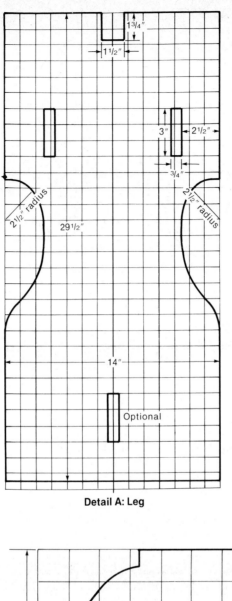

Detail A: Leg

With a handsaw or power saw, cut the plywood pieces. Cut the two uprights B as shown in the squared drawing in Detail A, using a saber saw or keyhole saw to shape the curved edges and the notch on the top edge. To start the slots for the stretchers in pieces B, make a plunge cut with a saber saw, or drill pilot holes and cut with a keyhole saw. Shape pieces D as shown in the squared drawing in Detail B. Cut the ends of stretchers C as shown in Detail C. Make the holes for the pegs (¾-inch square) by drilling through the stretchers, then squaring off the corners with a saber saw or keyhole saw.

STEP 2
Putting the bases on the uprights
Cut two 16½-inch lengths of 2 × 4 for base pieces E. Bevel the top ends with a plane or forming tool, then cut each length into two L shapes, as shown in Detail D. Glue base pieces E around the bottoms of uprights B, then secure them with countersunk No. 10 × 2-inch flat head wood screws, one on each side. Glue wood

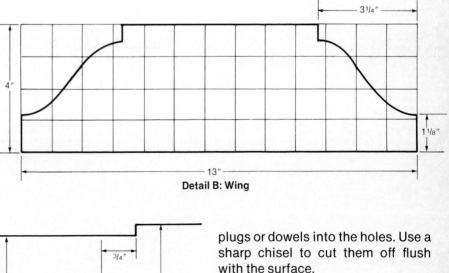

Detail B: Wing

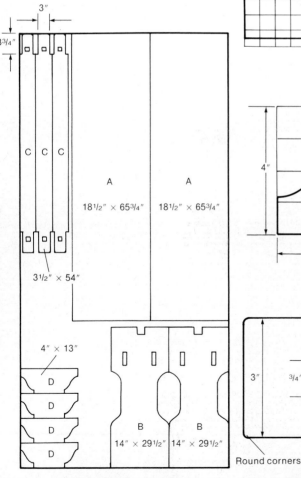

Panel Diagram Layout

Detail C: Stretcher Tongues

plugs or dowels into the holes. Use a sharp chisel to cut them off flush with the surface.

STEP 3
Attaching supports to uprights and finishing the table top
With a hacksaw, cut four 4-inch lengths of continuous hinge to attach pieces D to uprights B, as

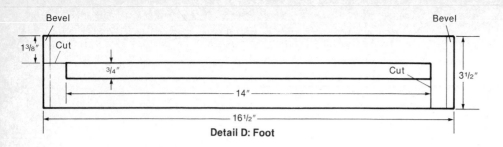

Detail D: Foot

shown in the Exploded View. You may also use a pair of 1-inch butt hinges to attach each piece D.

Cut two 66¾-inch lengths of 1 × 3 (F) and fasten them to the inner edges of table tops A with glue and brads, flush at the top and overhanging ½ inch at each end. Cut four 18½-inch lengths of ½ × ¾-inch trim (H); glue and nail them with brads to the ends of tops A. Cut two 66¾-inch lengths of ½ × ¾-inch trim (G) and fasten them to the outer edges of tops A. Cut four (or six, if you are using the third stretcher) 1 × 1 pegs (I), 3 inches long.

Finish all the pieces before assembly. Stain or varnish stain are recommended, although the table may also be painted or antiqued.

STEP 4
Assembling the table

To assemble the table, place stretchers C through the slots in uprights B and lock them in place by inserting pegs I into the holes, outside the uprights. Set the top assemblies on the uprights, with inner edges F fitting into the notches in the top edges of pieces B.

FIREPLACE BENCH

Materials List
* 1 sheet, 4′ × 8′ × ¾″ MDO plywood
  1½″ half round wood molding × 8′
  1½″ half round wood molding × 7′
  ¾″ half round molding × 4′
  ¾″ quarter round molding × 4′
  6d finishing nails
  3d finishing nails
  White glue
  Wood putty
* Semi-gloss enamel
  1 cushion, approximately 21″ × 44″

* Semi-gloss enamel is recommended as a finish, but if you prefer a different finish, A-B plywood may be used.

Tools
Handsaw or saber saw
Hammer
Nailset
Coping saw
Miter box
Putty knife
Sandpaper

Length of Time Required
Afternoon

Safety Precautions
Safety goggles

When the weather outside is frightful, this fireplace bench will provide comfortable seating before the blazing hearth. And, when the fire gets low, you won't have to brave the elements for a trip to the woodshed. There's plenty of room under the bench seat for wood, kindling, and old newspapers.

STEP 1
Cutting the parts and building the box
Carefully measure and mark the plywood as shown in the Panel Diagram. Use a handsaw or power saw to cut the plywood pieces (kerf is allowed for in the layout).

With glue and 3d finishing nails, fasten two parts F to each side A, flush at front and back and ¾ inch from both top and bottom. Use glue and 6d finishing nails to fasten top and bottom pieces B to the A-F end assemblies. Glue and nail partitions E between pieces B. With glue and 6d nails, attach back D to pieces B, E, and F. Glue and nail piece C across the front, flush with top B.

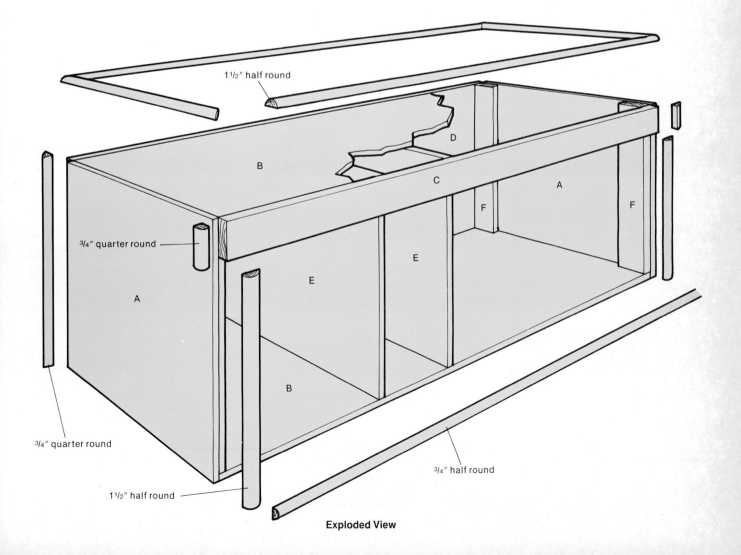

Exploded View

STEP 2
Adding trim

From the 7-foot length of 1½-inch half round molding, cut two 13¼-inch pieces. Use glue and 3d nails to fasten them to sides A, flush at the bottom and the outer edges. Show off your woodworking skill by coping the ¾-inch half round molding at the bottom front around the half round on the sides and attach to base B. Cut and fasten the ¾-inch quarter round molding between pieces A and C and pieces A and D, as shown in Detail A. Attach the remaining 1½-inch half round around the top edge of the bench, mitering at the corners.

STEP 3
Finishing the bench

Fill joints with wood putty as needed. Drive all visible nails below the surface with a nailset and fill the holes with wood putty. Sand smooth.

Apply semi-gloss enamel or other finish, as desired. When thoroughly dry, set a cushion in place, build a roaring blaze in the fireplace, settle in, and be mesmerized by the leaping flames.

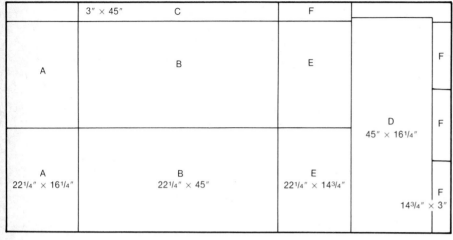

Panel Diagram

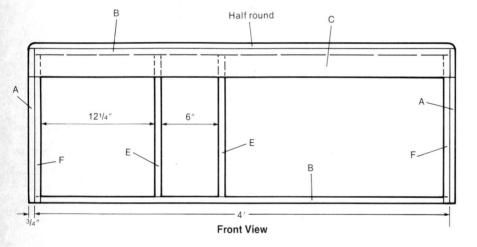

Front View

12¼″   6″

4′

¾″

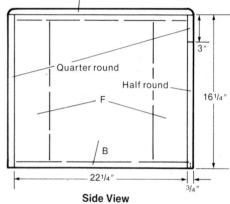

Side View

3″

16¼″

22¼″   ¾″

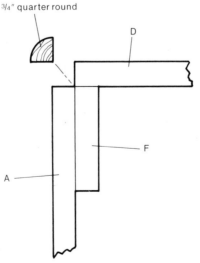

Detail A: Corner

¾″ quarter round

STORM WINDOW

Materials List
$^5/_4'' \times 2''$ lumber (fir, pine, spruce), length as required
$^1/_2''$ quarter round molding, length as required
$^3/_8''$ dowel $\times$ 4'
1" brads
Glass, to required size
Waterproof glue
Wood putty
Glazier's points
Glazier's compound
Linseed oil
Storm sash hardware
Exterior trim paint

Tools
Stationary power saw or handsaw
Vise or clamps
Square
Hand brace or power drill
Chisel
Back saw and miter box
Hammer
Nailset
Putty knife
Plane

Length of Time Required
Afternoon

Safety Precautions
Safety goggles

No one needs to be told that, with to-day's heating costs, it's more important than ever to button up your house tightly for the winter. You can buy stock-size storm sash for windows, but odd-size windows are another matter. You pay exorbitant prices for custom-made storm windows. So take matters into your own do-it-yourself hands, and put the money you save toward your next oil or gas bill.

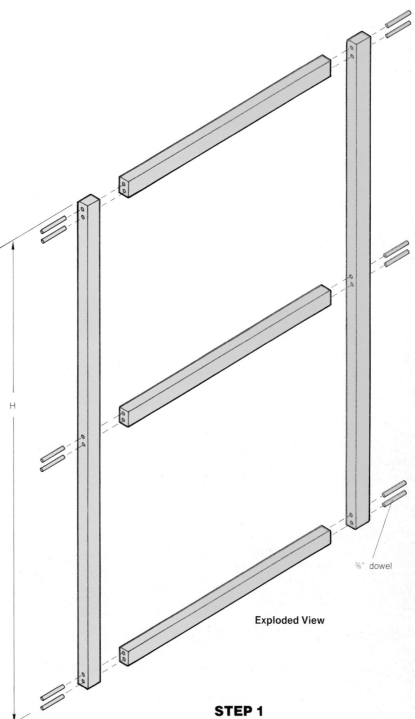

H

⅜" dowel

Exploded View

STEP 1
Measuring and cutting the parts
Measure the exact height and width just *inside* the outer edge of the exterior window frame. As shown in the Exploded View, you will need two

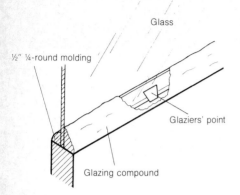

Glass

½" ¼-round molding

Glaziers' point

Glazing compound

pieces of ⁵/₄ × 2 stock (the nominal ⁵/₄ dimension is an actual 1¹/₁₆ inches and will be the thickness of the sash) the height of the frame and three pieces the width of the frame less 3 inches (the combined width of the vertical members). Cut these to exact height and width, making sure all cuts are perfectly square.

STEP 2
Assembling the frame
Cut twelve pieces of dowel, each approximately 3½ inches long. Slightly chamfer one end of each dowel with a piece of medium sandpaper. Hold the frame pieces in a vise or with clamps, making sure that they are perfectly square, and drill two ³/₈-inch holes approximately 3¹/₈ inches deep at each joint. Assemble the sash frame by coating the dowels with glue and driving the chamfered edges through the vertical members into the horizontals—first at top and bottom, then into the middle piece. Cut off protruding ends of the dowels with a sharp chisel or a fine-tooth saw and wipe away any excess glue.

STEP 3
Adding molding
Fasten the quarter round molding flush with the inside edge of the frame around both the top and bottom sections. Miter the corners, using a back saw and miter box. Use glue and 1-inch brads to secure the molding. Set the brads below the surface, fill the holes with wood putty, and sand smooth.

STEP 4
Adding the glass
Glass should be cut to fit the inside of the frame, less ¹/₁₆ inch in each direction. Coat the inside of the frame (where glazier's compound will be applied) with linseed oil, apply a thin coating of glazier's compound against the molding, and press the glass in place. Insert two glazier's points along each side to hold the glass snug against the molding. (Glazier's points usually come with a driving tool to minimize the likelihood of smashing the glass. Use it.) Apply glazier's compound all around the glass, flush with the outer edge of the frame and with the molding on the inside of the glass. Use a putty knife to smooth out the compound.

STEP 5
Fitting and finishing the window
Plane the bottom edge of the storm sash to match the slope of the window sill. Fasten sash hangers 2 inches in from each edge at the top, with the other half of each hanger fastened to the top of the window frame. Attach bottom sash hardware. Check the sash for fit. Plane, if necessary (but not so much that you create an air leak). After you are satisfied with the fit, remove the sash and apply trim paint, following manufacturer's directions.

PICNIC TABLE

Materials List
* Table top pieces (6), 2″ × 6″ × 95″
  Legs (6), 2″ × 4″ × 32″
  Seat braces (3), 2″ × 4″ × 55″
  Cleats (3), 2″ × 4″ × 30″
  Bench seats (4), 2″ × 6″ × 95″
  Angle braces (4), 2″ × 4″ × 20″
  No. 8 × 2″, No. 8 × 2½″, No. 8 × 3½″
  flat head wood screws
  No. 8 flat head galvanized nails
  (optional)

4½″ lag bolts
Paint or stain

Tools
Portable circular saw or handsaw
Carpenter's square
Drill
Screwdriver
Hammer (optional)
Power sander or sandpaper

Length of Time Required
Afternoon

Safety Precautions
Safety goggles and gloves
* All lumber either construction-grade stock or
  heartwood redwood.

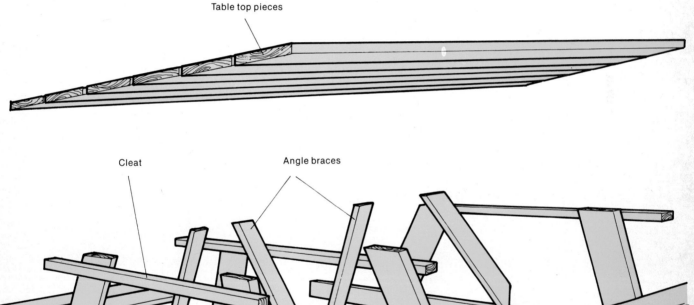

Table top pieces

Cleat

Angle braces

Leg

Bench seats

Seat braces

Exploded View

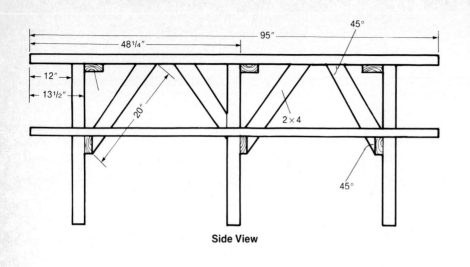

Side View

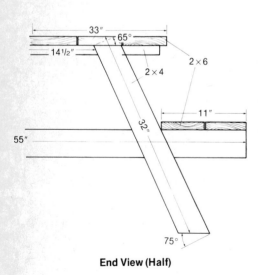

End View (Half)

A picnic table is one of the most useful outdoor items you can build. Not only is it practical, but the table is fun to build and fairly economical if you use standard construction-grade 2×4s and 2×6s. In fact, the table shown here was made from 2×4 and 2×6 scrap. The finished table was given a coat of redwood stain. It could also be painted with a bright acrylic paint. Heartwood redwood makes a longer lasting if more expensive table that will not require any maintenance or finish, as does a table made from construction-grade fir or pine.

One disadvantage to using construction-grade material is knots, and you may have some waste if you have to cut around many. However, many building supply yards will allow you to pick material from a stack, especially if you tell them what you're building. If the material does have knots, make sure they are solid and sound, not loose, and with no holes or cracks in them. You should also make sure the stock is straight —avoid warped or twisted stock. The table can be constructed using hand tools or using a portable circular saw, sander, and drill.

STEP 1
Cutting the pieces
Cut all the pieces for the table top and bench seats to the given dimensions. The easiest way to do this is to lay out the lumber across two sawhorses and mark the cuts with a carpenter's square. Make the cuts with a portable circular saw or a handsaw. Follow the same procedure to lay out and cut the seat braces and top cleats. Then lay out the angled braces and the legs and mark them for length and the given angles (see the Side and End views). Cut to size. Lightly sand all pieces, taking care to sand away any splinters and rough edges on the cut ends.

STEP 2
Assembling the table top
Lay out the top pieces upside down on a flat surface (the garage floor is fine), square up the ends, and butt the side edges together. Fasten the three cleats across the top pieces in the position shown (see the Side View) with No. 8 × 2-inch flat head wood screws driven through the cleats into the top pieces. You could also use No. 8 flat head galvanized nails to fasten the cleats.

STEP 3
Assembling the leg sections
Lay out the legs on a flat surface, against a straightedge, with the bottom inside corners 40 inches apart. Position the seat braces horizontally across the legs so that they overhang the legs by 11¼ inches on their top edges at either end (about half way up the legs). Mark the position of the seat brace–leg intersection so you can return the braces to position if they slip, then fasten the braces to the legs using countersunk No. 8 × 2-inch wood screws or No. 8 flat head galvanized nails. Assemble all three leg sections in this manner.

STEP 4
Attaching the legs to the table top
With the table top face down on a flat surface, attach the legs to the cleats with 4½-inch lag bolts in the positions shown in the Side View. Then fasten the angled braces as shown in the Side View with countersunk No. 8 × 2½-inch and No. 8 × 3½-inch flat head wood screws.

STEP 5
Installing the bench seats
Turn the table right side up and attach the bench seats to the seat braces with countersunk No. 8 × 2-inch wood screws, making sure that no burrs stick up from the screws to snag clothes. Sand the entire table smooth and finish.

BARBECUE CART

Materials List

Back (1), 24" × 34½" × ¾" A-B plywood
Sides (2), 19¼" × 24" × ¾" A-B plywood
Top (1), 20" × 36" × ¾" plywood
Bottom and top facers (1 each), ¾" × 2" × 36" white pine
Side facers (2), ¾" × 2" × 20" white pine
Bottom (1), 18½" × 34½" × ¾" plywood
Doors (2), 16⅜" × 20¾" × ¾" plywood
Bottom cleats (2), 1" × 34" × ¾" plywood
Bottom cleats (2), 1" × 16" × ¾" plywood
Side rails (2), 2½" × 48" × ¾" plywood
End rails (2), 1½" × 20" × ¾" plywood
Handle (1), 1" dowel × 23"
Chopping block pieces (26), ¾" × 1½" × 12" maple
Rear legs (2), 11" × 2" × 6" lumber
Front legs (2), 8" × 2" × 6" lumber
Axle (1), ½" threaded rod, cut to length
6" lawnmower wheels (2)
Ceramic tile for 24" × 20" surface
Nuts and washers for axle
⅜" lip hinges (4)
Cabinet door knobs (2)
Magnetic catches (2)
No. 8 × 1½" round head wood screws
8d finishing nails
6d finishing nails
White glue
Waterproof glue
Resorcinol glue
Type 1 mastic
Silicone grout finish
Wood putty
Exterior enamel paint

Tools

Table saw or radial arm saw
Band saw or saber saw
Framing square
Hacksaw
Wrench
Screwdriver
Power drill, ½" and 1" bits
Hammer
Nailset
Router or shaper with quarter round or bead cutter
Power sander or sandpaper
Bar clamps
Notched trowel

Length of Time Required

A few days

Safety Precautions

Safety goggles

Most barbecue chefs ask themselves the same question: Where's the best place to store bags of charcoal, lighter fluid, and outdoor cooking utensils? And just as the steaks start to flare up on the outdoor grill, most chefs look for a place to set down a drink or the marinade so they can tend to the cooking. This barbecue cart takes care of all those details and then some. It not only has storage space for charcoal, paper plates, glasses, and other items needed for cookouts. It has hanging space on the ends for barbecue utensils. A portion of the top is covered with a maple cutting board that makes easy work of trimming meat and shaping hamburgers. The rest of the top is covered with easy-to-clean ceramic tile. Hot dishes and items can be set on the tile top. The cart is equipped with wheels so hot-off-the-grill meals can easily be rolled to a serving table. Except for a few pieces, the cart is constructed of Exterior plywood and fastened together with nails and waterproof glue. It is then given several coats of good-quality exterior enamel so it can withstand the weathering abuse of being left outdoors.

STEP 1
Cutting and assembling the shell

Cut the back and sides to the given dimensions from ¾-inch Exterior

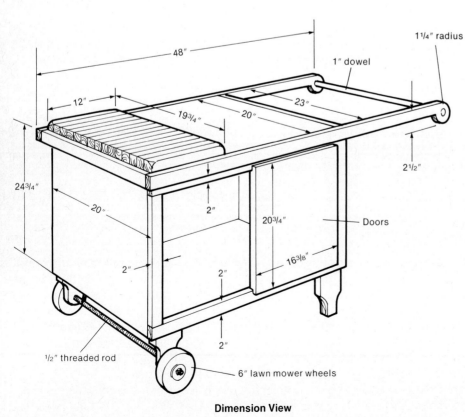

Dimension View

plywood. Then fasten the back to the sides, as shown here, using 8d finishing nails and waterproof glue. Set all nails and fill with wood putty as you work. Cut the top and bottom facer strips from ³/₄-inch white pine. Turn the plywood shell over on its back on a smooth, flat surface and fasten the facer strips in place between the side pieces, flush with the top and bottom edges, using glue and 8d finishing nails. Again, set the nails and fill with wood putty. Cut the side facers from ³/₄-inch white pine; glue and nail in place. Drive 8d finishing nails up from the underside of bottom facers into the ends of the side facers. Drive nails down from the top of the top facers into the ends of the side facers. Cut the bottom cleats to the given dimensions and glue and nail them in place around the inside edge of the bottom, ³/₄ inch up from the edge. Cut the ³/₄-inch plywood bottom to size, and fasten it in place on the cleats with glue and 6d finishing nails.

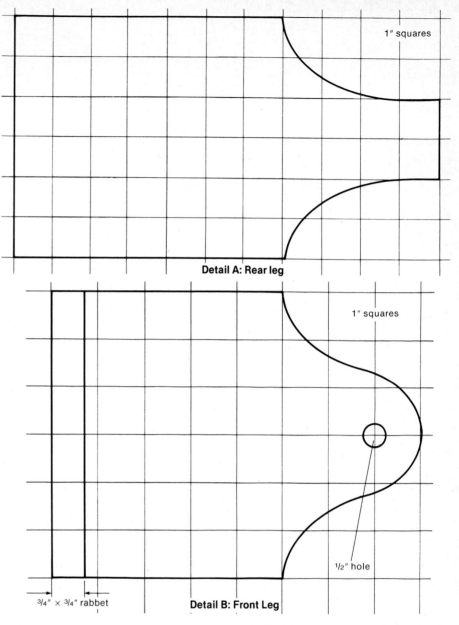

Detail A: Rear leg

1″ squares

1″ squares

¹/₂″ hole

³/₄″ × ³/₄″ rabbet   Detail B: Front Leg

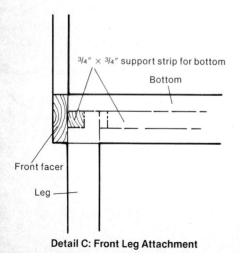

Adding bottom facer.

³/₄″ × ³/₄″ support strip for bottom

Bottom

Front facer

Leg

Detail C: Front Leg Attachment

STEP 2
Adding the legs
Enlarge the squared drawings for the wheel supports and legs (Details A and B). Transfer these patterns to blocks of 2×6 and cut the ³/₄ × ³/₄-inch rabbets on the ends to allow the blocks to fit over the cleats. Then use a band saw or a saber saw to cut out the profiles of the pieces. Bore the ¹/₂-inch holes for the threaded rod axle in the ends of the wheel supports. Sand smooth, then fasten in place to the bottom of the cart (see Detail C), using glue and No. 8 × 1¹/₂-inch round head wood screws. Insert the rod in place, but don't cut it to length yet. Place washers and one lawnmower wheel on one end of the axle and the other wheel on the long

end; hold the wheels securely in place with nuts. Use a hacksaw to cut the excess rod off *one* end of the axle.

STEP 3
Adding the top and rails
Turn the cart right side up. Enlarge the squared drawing (Detail D) for the side rails. Cut the rails to shape on a band saw, smooth all edges, and bore the 1-inch hole in the center of each round end for the dowel handle. Cut a piece of ³/₄-inch plywood to the given dimensions for the top and fasten it in place with glue and 6d finishing nails.

Lay a piece of the ceramic tile you intend to use in place on one edge of the top of the cart. Then position a

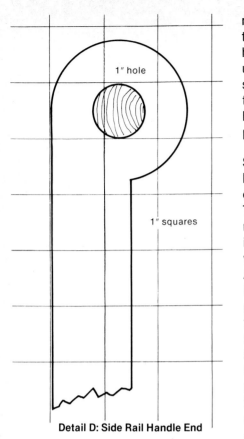

1" hole

1" squares

Detail D: Side Rail Handle End

side rail against the tile. Adjust so that the tile protrudes slightly above the top edge of the rail. Mark the height of the rail along the lower edge on the side of the cart. Then install the side and end rails with glue and 6d finishing nails in the position you have marked. Glue the handle between the two side rails.

STEP 4
Cutting the doors

Cut the doors from 3/4-inch Exterior plywood to the given dimensions. Round the outside edges of the doors with a quarter round or a bead cutter in a router or shaper. Cut 3/8-inch rabbets in the inside edges at the top, bottom, and one side with a table saw or with a radial arm saw and dado blade.

STEP 5
Painting the cart and adding doors and shelves

At this point you should completely sand the entire cart, including the doors, and paint or finish as desired. Use several coats of the finish you choose. After the finish has dried, install the hinges and knobs and hang the cabinet doors in place. Use mag-

netic catches on the inside to hold the doors securely shut. Screw cup hooks onto the ends for cooking utensils. You could also install a shelf in the cabinet (use cleats, as on the bottom) to create even more usable space for plates and other patio party items.

STEP 6
Making and installing the cutting board

The maple cutting board is made of maple strips, 1½ inches wide and ¾ inch thick, cut to length and ripped to width; the top edges are then jointed. After cutting and jointing all pieces, lay them in position on clean newspaper placed on a flat, smooth surface. Mix up waterproof two-part resorcinol glue and spread it on both surfaces of each side of the strips. Position the strips together on edge, use a framing square to make sure they are formed into a square block, and clamp solidly with bar clamps. Allow to dry overnight. Remove clamps and sand the block as smooth as possible to even out the surfaces of the strips. Round the top edge with a bead cutter in a shaper. Use waterproof glue to glue the finished board to the cart top at the end opposite the handles.

STEP 7
Adding the tile

Use a notched trowel to spread a good water-resistant mastic (Type 1) for ceramic tile on the area of the cart top not covered by the cutting board. Position the tile in place and press it down firmly. Don't move it around too much—you will squeeze out too much adhesive. Allow the mastic to set overnight. Grout, using a latex-based grout, and wipe away excess with a sponge. Wait about fifteen minutes. Clean away any other excess grout with a damp cloth. Wait about thirty minutes, then polish the tile surface with a clean cloth and apply a silicone grout finish for further protection.

Now all that's needed is to call in the kids and neighbors for a barbecue.

Installing hinges on the doors.

BLANKET CHEST

Materials List

Sides (2), $3/4'' \times 14'' \times 19^1/2''$, glued-up
   stock
Front and back (2), $3/4'' \times 14'' \times 30''$
   glued-up stock
Bottom (1), $3/4'' \times 18'' \times 28^1/2''$ plywood
Top (1), $3/4'' \times 20^1/2'' \times 31''$ glued-up
   stock
Cross braces (2), $1'' \times 3'' \times 18''$
Corner blocks (4), $1'' \times 1'' \times 12^3/4''$
Flat-bottomed casters (4), with screws
Blanket hinges (2), with screws
Glue
No. $8 \times 1^1/4''$, No. $8 \times 1^1/2''$ flat head
   wood screws
8d finishing nails
Stain or clear finish

Tools

Handsaw, table saw, or radial arm saw
Hand drill or power drill
Doweling jig
Bar clamps
Screwdriver
Hammer
Sander or sanding block, sandpaper

Length of Time Required
Overnight

Safety Precautions
Safety goggles

A large, sturdy box with a lid is some-thing every crafts-oriented parent should build for his or her children. Kids like to have a place to put away toys, and parents like them to have one, too. This chest is also useful outside the playroom or children's bedroom; it makes an excellent place to store blankets or linens. The one shown here was made from red cedar stock, a fragrant wood that im-parts a fresh odor to blankets and lin-ens. Cedar is also reputed to repel moths, so it is good for storing woolen items. Kids aren't much inter-ested in blanket storage, but if you make one chest for them and an-other for yourself, they won't mind.

The chest shown here happens to be one I made entirely with hand tools, and I recommend it for polish-ing handwork skills. I completed the chest with a handsaw, hand drill, doweling jig, bar clamp, hammer, and screwdriver.

STEP 1
Doweling the top and sides
The dimensions for the chest are ar-bitrary; you can vary them to build a chest whose size depends on your needs or the materials you have on hand. If you use your own dimen-sions, just make sure that the oppos-ing sides are exactly the same size, and when the box is assembled, measure its inside dimensions for fit-ting the bottom, and its outside di-mensions for fitting the lid. The chest sides and lid are made of glued and

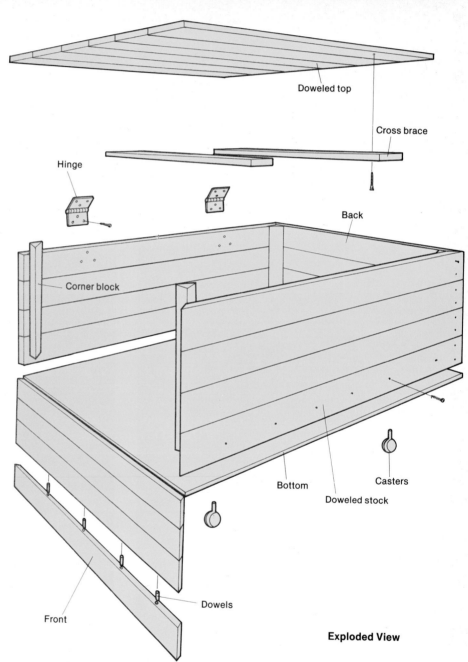

Doweled top

Cross brace

Hinge

Back

Corner block

Bottom

Casters

Doweled stock

Dowels

Front

Exploded View

doweled stock; the bottom is ¾-inch plywood.

To glue and dowel the stock for each side and the lid, lay out pieces of ¾-inch stock, alternating the direction of the grain in every other piece to minimize warpage, until you have enough stock to make a side. Smooth the butted edges of the stock with a jointer (or a hand plane) until the pieces fit perfectly against each other. Mark four parallel lines across the width of the pieces (at a right angle to the butted edges) so that they divide the stock into five equal parts. Use a doweling jig to drill a 1-inch-deep hole at the four marks in each facing edge. Squeeze glue into the holes along the edge of one piece of stock and drive in 1⅜-inch spiral dowels; glue along the facing edge of the next piece and drive it onto the dowels. Repeat the procedure until you have glued and doweled all the stock for the piece you are making, then fix the stock in a bar clamp and tighten the clamp until the glue squeezes out of the joint lines. Wipe the glue away with a warm, damp cloth. Allow to dry overnight. Use this same procedure for making the other sides and lid.

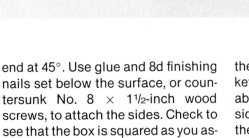

STEP 2
Cutting the sides and assembling the box
When the pieces are dry, cut each side to the proper size and miter each

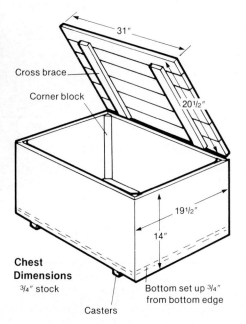

Chest Dimensions
¾" stock
31"
20½"
19½"
14"
Cross brace
Corner block
Casters
Bottom set up ¾" from bottom edge

end at 45°. Use glue and 8d finishing nails set below the surface, or countersunk No. 8 × 1½-inch wood screws, to attach the sides. Check to see that the box is squared as you assemble the pieces.

STEP 3
Cutting and fitting the corner blocks and bottom
Cut the bottom from ¾-inch plywood to the inside dimensions of the box you are building. Rip 45° corner blocks 1¼ inches shorter than the depth of the box and miter the tops at 45°. Install them with glue. Attach the bottom with glue and No. 8 × 1½-inch wood screws driven in through the sides.

STEP 4
Building the lid
Cut the stock you've glued and the dowel for the lid to 1 inch larger than

the outside dimensions of your blanket chest. Cut the cross braces to about 2 inches shorter than the inside width of the box; they must clear the edges for the lid to close. Attach the braces across the lid with countersunk No. 8 × 1¼-inch flat head wood screws.

STEP 5
Finishing the chest
Position the lid so that it overlaps the edges by the same distance on all sides. Attach it with standard blanket chest hinges, which are available at hardware stores or home centers. Attach flat-bottomed casters to the bottom of the box at the corners. Finish the chest however you like, inside and out. If you use cedar stock, as I did, don't finish the inside: if you do, you will seal in the oils that give the wood its pleasant smell.

BUTLER'S TRAY TABLE

Materials List
1 sheet, 4' × 8' × ½" A-B plywood
1 piece, 2" × 2" × 8' lumber
1 piece, 1" × 2" × 6' lumber
1 piece, ¼" × 1" × 6' wood strip
Spring-lock hinges (Brass Hinges No.
   12004, from Minnesota Woodworkers
    Supply Co. are shown) (4)
Furniture glides (4)
Glue (aliphatic resin, urea formaldehyde)
4d finishing nails
Wood putty
Stain
Clear finish

Tools
Saber saw or coping saw
Stationary power saw or handsaw
Hand brace or power drill
Chisel
Screwdriver
Hammer
C clamps and bar, pipe, or web clamps
Sanding block or power sander

Length of Time Required
Overnight

Safety Precautions
Safety goggles

This traditional English tea table has a top that is actually a removable tray, with hand openings for easy carrying and sides that flip up on spring-lock hinges.

STEP 1
Cutting the tray pieces
Mark the plywood as shown in the Panel Layout; consult the Top View for laying out tray pieces A, B, and C. Cut out the oval tray as a single piece, without separating ends B and sides C from top A. To cut the hand openings in B and C, drill 1-inch holes at each end, then cut between them with a saber saw or coping saw. Cut ends B and sides C from top A.

STEP 2
Completing the tray
With a sharp pencil, trace the outlines of the hinges on top A and ends B, 3 inches from the corners. Trace the hinge outlines on top A and sides C, 4 inches in from the corners. With a sharp chisel, outline the mortise cuts for the hinges, then clean out the plywood veneers—a thin sliver at a time—until the mortise is just deep enough to accommodate each hinge flush with the surface. (If you remove too much wood, just fill in with thin scraps to bring the hinge up to level.) Attach ends B and sides C to top A with the hinges.

On the underside of A, mark the locations of guides G, as shown in the Top View. Cut two 1 × 2 guides, each 26 inches long and angled 60° at each end (see the Exploded View). Glue the guides to the bottom of the tray.

½" slot cut diagonally through leg corner to corner

3"

Corners of bottom cut diagonally to fit into leg slots

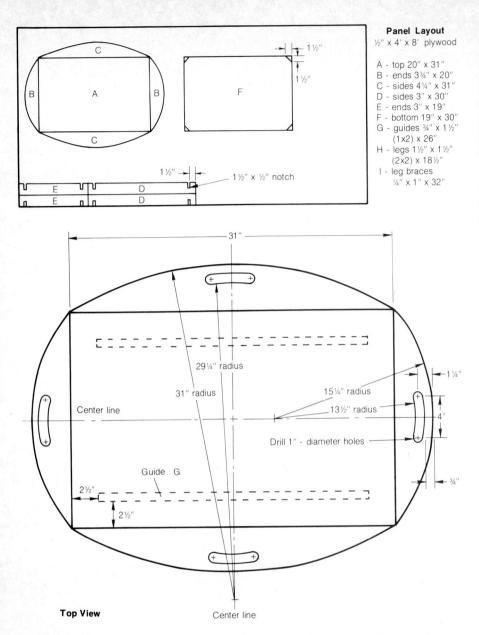

Panel Layout
½" x 4' x 8' plywood

A - top 20" x 31"
B - ends 3¾" x 20"
C - sides 4¼" x 31"
D - sides 3" x 30"
E - ends 3" x 19"
F - bottom 19" x 30"
G - guides ¾" x 1½"
 (1x2) x 26"
H - legs 1½" x 1½"
 (2x2) x 18½"
I - leg braces
 ¼" x 1" x 32"

1½" x ½" notch

31"

29¼" radius

31" radius

Center line

15¼" radius

13½" radius

1¼"

4"

Drill 1" - diameter holes

¾"

Guide G

2½"

2½"

Top View

Center line

next, and clean out the waste with a chisel. Check the fit by assembling the bottom unit and the legs without glue. Make sure everything is square. Measure diagonally between legs for the exact lengths of the ¼ × 1-inch leg braces (approximately 32½ inches). Cut the braces (I) and notch them where they will cross in the center (see Detail A).

STEP 5
Completing the assembly
Apply glue to the inside tops of the legs and the corners of bottom F and fasten the legs to the bottom unit. Apply glue to the ends of braces I and the notch in the center, then insert pieces I into the mortises in the legs. Nail through sides and ends D and E into legs H. Clamp the leg assembly together with bar or pipe clamps or a nylon web clamp. Use a C clamp to hold the two pieces I securely together at the center. Wipe away any excess glue that is squeezed out of the joints. Allow to dry thoroughly before removing clamps.

STEP 6
Finishing the table
Set all nails below the surface and fill the holes with wood putty. Fill any raw plywood edges with wood putty (or use wood tape to cover the edges). Sand smooth. Apply stain to both table unit and tray, taking care to wipe stain off the hinges. Apply a clear finish, such as polyurethane varnish, to protect the new piece of furniture. Tap the glides onto the leg bottoms to complete the project.

STEP 3
Assembling the table
Cut out table bottom F, with diagonal corner cuts as shown in the Panel Layout. Cut sides D and ends E, notching them as shown in the Panel Layout. Assemble sides and ends D and E with glue, fitting the notches together. Glue assembly D–E to bottom F and secure by driving nails through F into D and E. Clamp until glue dries, and wipe away any excess glue that squeezes out of the joints.

STEP 4
Cutting the legs
Cut four 2×2 legs (H), each 18½ inches long. Cut a ½-inch diagonal slot from corner to corner, 3 inches

from the top of each leg (see the Exploded View). Make a mortise ¼ inch deep and 1 inch long into the inside corner of each leg, 3½ inches from the bottom. To do this, drill four ¼-inch holes, each just touching the

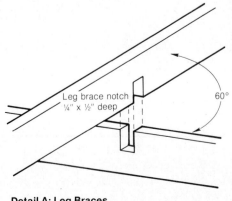

Leg brace notch
¼" x ½" deep

60°

Detail A: Leg Braces

COMPACT DESK

Materials List

1 sheet, 4' × 8' × ¾" A-B plywood
1½" half round molding × 16'
¾" half round molding × 14'
* 1 yard of vinyl material
3d finishing nails
8d finishing nails
1½" brads
1" brads
† Vinyl adhesive
White glue
Wood putty
Semi-gloss enamel

* Available at wallcovering and many department stores.
† Purchase adhesive as recommended by vinyl material manufacturer.

Tools

Stationary power saw, circular saw, or handsaw
Saber saw or keyhole saw
Hand brace or power drill
Hammer
Nailset
Roller or rolling pin
Mat knife
Clamps
Sanding block

Length of Time Required

Overnight

Safety Precautions

Safety goggles

Ideal for the student's room or the home office, this desk measures just 2 feet deep and a little over 3 feet wide; yet it offers a spacious, vinyl-covered work surface and ample shelf space underneath for supplies. It is also a sturdy, handsome piece of contemporary furniture.

STEP 1
Cutting the parts

Carefully measure and mark the plywood as shown in the Panel Layout. With a handsaw or a power saw, cut pieces B, C, D, G, H, and I. To start the inside corner cuts on pieces A, make a plunge cut with a saber saw, or drill a pilot hole and cut with a keyhole saw; then square the corners. Cut out pieces A, E, and F.

STEP 2
Making the sides

The two pairs of pieces A are glued and nailed together to form the desk sides. Make sure the good faces of the plywood are facing out. Apply a bead of glue around the edges and down the middle, then fasten with 3d finishing nails, taking care not to deface the plywood with the hammer. Finish driving the nails beneath the surface with a nailset. Clamp the pieces together (or place weights on them) until the glue has dried.

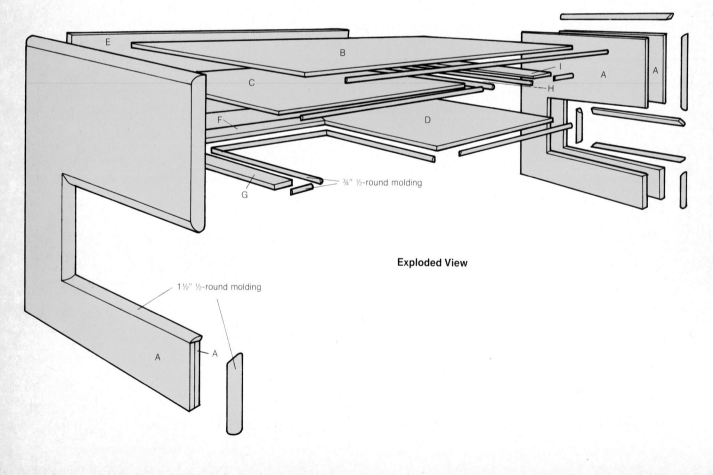

Exploded View

¾" ½-round molding

1½" ½-round molding

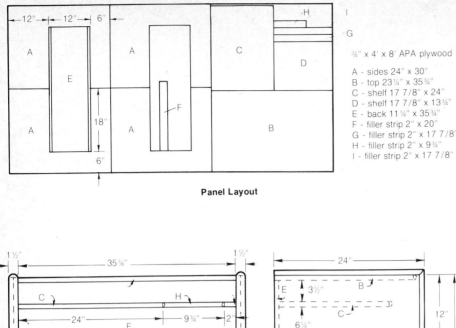

³/₄" x 4' x 8' APA plywood

A - sides 24" x 30"
B - top 23¼" x 35¾"
C - shelf 17 7/8" x 24"
D - shelf 17 7/8" x 13¾"
E - back 11¼" x 35¾"
F - filler strip 2" x 20"
G - filler strip 2" x 17 7/8"
H - filler strip 2" x 9¾"
I - filler strip 2" x 17 7/8"

Panel Layout

Front View

Side View

All edges of sides A except the back and bottom are covered by 1½-inch half round molding (see the Side and Exploded views). Cut the molding in a miter box with a back saw, making 45° miters where the molding pieces meet. Glue the molding to the edges of the two sides, then secure them with 1½-inch brads.

STEP 3
Finishing the desk top
Cut a piece of ³/₄-inch half round molding to fit the front edge of piece B (the desk top). Glue it in place and secure with 1-inch brads. Wipe piece B clean of all dirt and dust, then apply vinyl adhesive to the top (the good side), the back edge, the front molding, and part way back on the underside. Press the vinyl onto the surface with a roller or rolling pin, rolling or pushing from the center toward the edges to remove any air bubbles. Allow the adhesive to dry. Use a mat knife to trim the vinyl flush with the bottom of the back edge and with the side edges of A.

STEP 4
Assembling the desk
Assemble the desk with glue and 8d finishing nails. Refer to the Exploded, Front, and Side views for placement of pieces. Fasten back E between sides A, ³/₄ inch below the top edges of sides A. Add shelf D, filler G, and filler F flush with the bottom of back E, nailing into them through the sides and back. Apply ³/₄-inch half round molding to all edges of D, G, and F, mitering at inside and outside corners. Add shelf C and fillers I and H and apply half round molding to exposed edges as above. Set top B in place and secure it by nailing through sides A.

STEP 5
Finishing the desk
Fill all exposed nail and brad holes (and any less-than-perfect miter joints) with wood putty. Sand smooth. Apply at least two coats of enamel. When it's dry, your desk is ready to go to work.

RUSTIC PLANTER

Materials List
* 1 piece, 2" × 12" × 4' redwood
  1 piece, 2" × 6" × 5' redwood
* 1 piece, 2" × 4" × 4' redwood
  20 No. 16 × 3½" flat head galvanized
  screws
16d common nails
60d common nails
Waterproof glue

Optional
1 piece, 2" × 4" × 1' redwood
No. 16 × 3" flat head galvanized screws
   (16)
2" swivel casters (4)
* The 4' measurement is exact; purchase longer
  pieces if you want to allow for error.

Tools
Stationary power saw, circular saw, or
   handsaw
Hand brace or power drill
Carpenter's square or combination
   square
Screwdriver
Hammer

Length of Time Required
Afternoon

Safety Precautions
Safety goggles

This planter is an easy, rewarding
project for the beginning woodworker. It looks good, and it will last a
lifetime.

Redwood is the best choice for
outdoor planter boxes because of its
resistance to decay, warping, insects, and the elements. It requires
no finish and weathers to a mellow
silvery gray. The planter shown here
is easy to build, and its rustic appearance will complement any porch,
patio, or deck. If you wish to make it
mobile, just add casters to the base.

STEP 1
Cutting the parts
Cut the lumber to length as follows,
taking care to keep all ends square:
for the sides—two pieces, each 2 ×
12 × 13½ inches, and two pieces,

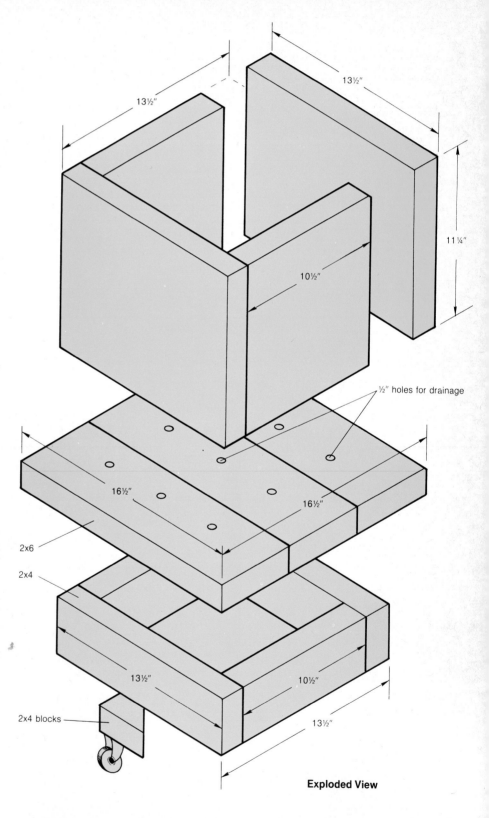

Exploded View

each 2 × 4 × 13½ inches; for the ends—two pieces, each 2 × 12 × 10½ inches; for the bottom—three pieces, each 2 × 6 × 16½ inches; and for the base ends—two pieces, each 2 × 4 × 10½ inches.

STEP 2
Assembling the planter box
Assemble the side ends of the planter with glue and 3½-inch screws, three at each corner. Pre-drill holes for the screws, 1 inch from both top and bottom, the third midway between. Be sure to keep the assembly square; check with a carpenter's or combination square.

Center the middle 2×6 bottom piece on the planter ends; fasten with glue and two 16d nails at each end. Add bottom pieces at each side of the center board, nailing to the planter ends and sides. Drill ½-inch drainage holes through the bottom boards.

STEP 3
Assembling and attaching the base
Assemble the base as you did the planter sides and ends, using only two screws at each corner. Drill ¼-inch holes up through the base (three on each side), then apply glue and drive 60d nails through the holes and the planter bottom into the planter sides. (Make sure these nails are not driven into the nails holding the bottom boards to the planter sides.)

STEP 4
Adding casters
If you wish to add casters to your planter, cut two lengths of 2 × 4, each 3½ inches long, then cut the pieces diagonally corner-to-corner. Fasten these blocks with glue and 3-inch screws driven through pre-drilled holes inside the corners of the base. Screw or bolt the casters to these blocks.

BOOKCASE

Materials List

Sides (2), $3/4'' \times 13 1/4'' \times 54''$
Top (1), $3/4'' \times 7'' \times 28 1/4''$
Top trim, sides (2), $1/2'' \times 2 1/4'' \times 8 1/4''$
Top trim, front and back (2), $1/2'' \times 2 1/4''$
  $\times 30 5/8''$
Back (1), $1/8'' \times 28 1/8'' \times 54''$
Front facer, top (1), $3/4'' \times 2'' \times 29 3/4''$
Side facer, top (2), $3/4'' \times 2'' \times 35 1/4''$
Shelves (3), $3/4'' \times 6 3/4'' \times 28''$
Storage top (1), $3/4'' \times 14 1/4'' \times 31''$
Facer, bottom (1), $3/4'' \times 2'' \times 29 3/4''$
Side facer, bottom (2), $3/4'' \times 2'' \times 11''$
Bottom door facer (1), $3/4'' \times 2'' \times 29 3/4''$
Bottom trim, front (1), $1/2'' \times 2'' \times 30 3/4''$
Bottom trim, sides (2), $1/2'' \times 2'' \times 14 1/2''$
Doors (2), $3/4'' \times 11 3/4'' \times 13 1/4''$
Cabinet hinges (4), $3/4''$ lip
Knobs (2)
Shelf hooks
Glue
4d flat head nails
6d, 8d finishing nails
Wood putty
Magnetic catches (2)
Stain or other finish

Tools

Circular saw or table saw
Handsaw
Electric drill or drill press
Router (optional)
Hammer
Screwdriver
Sander or sandpaper
Shaper
Carpenter's square
Plane or jointer
Nailset

Length of Time Required
Overnight

Safety Precautions,
Safety goggles, filter mask

A bookcase is a necessity for school-age youngsters, and it is one of the pieces of furniture they are likely to take along when they go off on their own. To withstand long use, this one is designed to be as sturdy and simple as possible. The enclosed bottom section is ideal for children's games and toys; the upper section has adjustable shelves to accommodate objects of various sizes, as well as books. The height of the entire unit is just right for school children. The bookcase shown has sides, top, bottom, and doors made of $3/4$-inch hardwood-faced oak plywood with the shelves, facing, and upper and lower banding made of solid oak. The back is constructed from $1/4$-inch paneling that happened to be left over from another project; you can use any comparable $1/4$-inch material that is handy.

STEP 1
Cutting and boring the sides
For the two side pieces shown in Detail A, mark off the dimensions on hardwood-faced plywood and cut them out with a circular or table saw. With both pieces, finish the inside cut in the L with a handsaw to get a

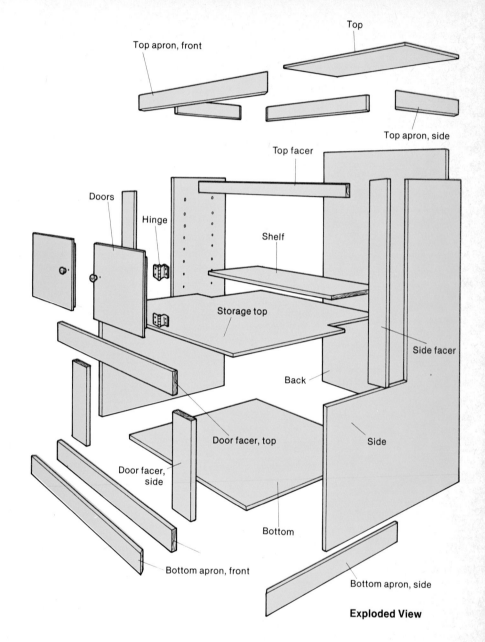

Top apron, front

Top

Top apron, side

Top facer

Doors

Hinge

Shelf

Storage top

Side facer

Back

Door facer, top

Side

Door facer, side

Bottom

Bottom apron, front

Bottom apron, side

Exploded View

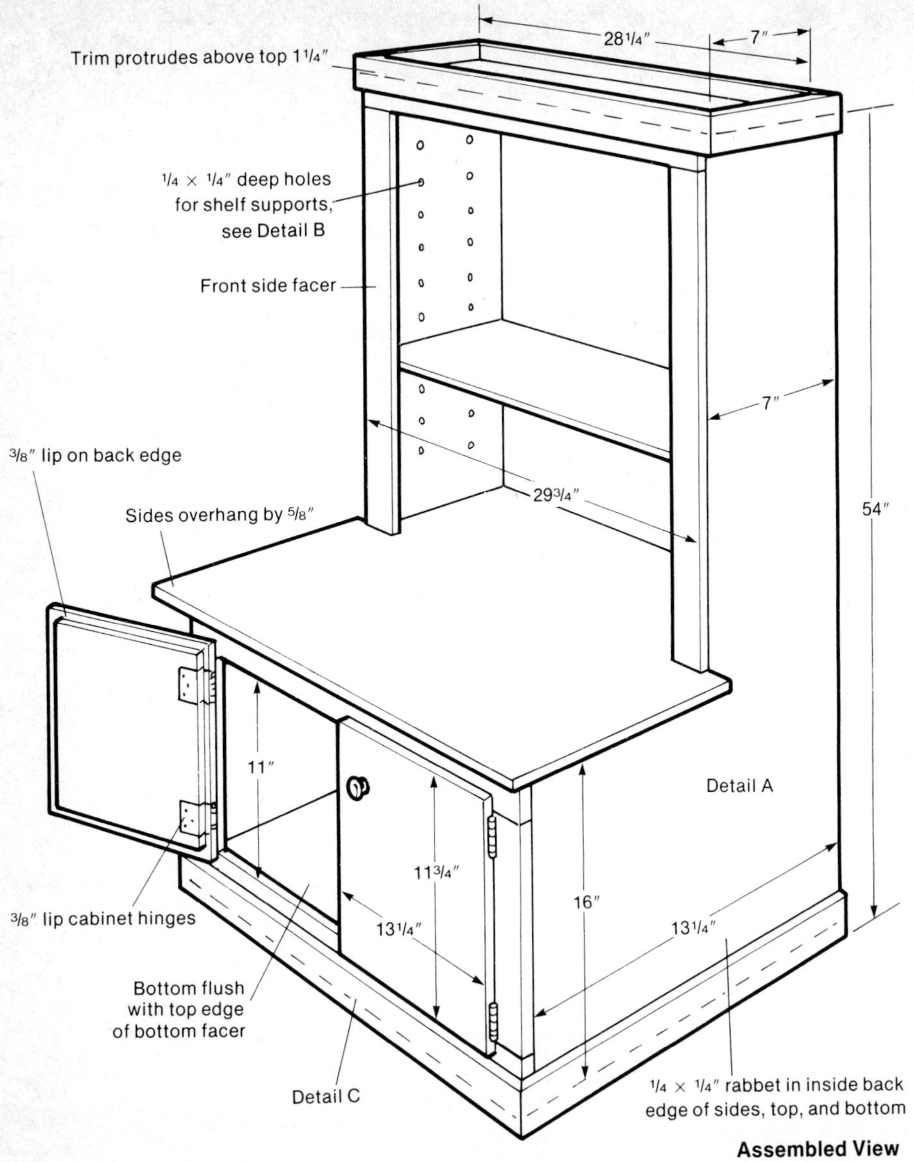
Trim protrudes above top 1 1/4"
28 1/4"
7"
1/4 × 1/4" deep holes
for shelf supports,
see Detail B
Front side facer
7"
3/8" lip on back edge
29 3/4"
Sides overhang by 5/8"
54"
11"
Detail A
11 3/4"
16"
3/8" lip cabinet hinges
13 1/4"
13 1/4"
Bottom flush
with top edge
of bottom facer
Detail C
1/4 × 1/4" rabbet in inside back
edge of sides, top, and bottom

Assembled View

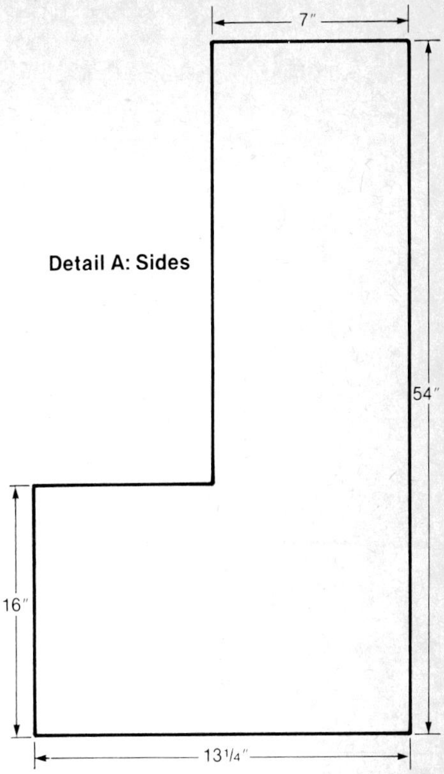
7"
54"
16"
13 1/4"

Detail A: Sides

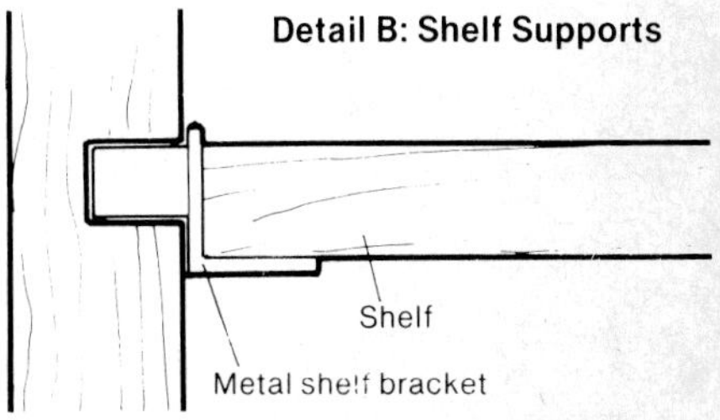
Shelf
Metal shelf bracket

Detail B: Shelf Supports

perfect corner. Sand all the edges smooth and cut a 1/4 × 1/4-inch rabbet in the inside back edge of each piece (check the Exploded View for location). The shelves will sit on small metal brackets plugged into stopped holes in the upper sides as shown in Detail B: there are two at each end of each shelf, and you should put them wherever else you think you might need them, or even all the way up the side, spaced at regular intervals, starting from the minimum height you want the first shelf. To mark the sides for drilling, check to make sure they are square, lay them out with the two long edges together, measure down from the top, and mark across the sides perpendicular to the long edge. Mark the points for the stopped holes 1 1/2 inches in from edges. Drill 1/4-inch holes 3/8 inches deep with a portable electric drill with a drill stop or use a drill press.

STEP 2
Assembling the frame

Cut the top and bottom to size and also cut a 1/4 × 1/4-inch rabbet on their inside back edges. Assemble the sides and top (between the sides, flush with the top edges) and bottom (set 3 inches up from the bottom of the side), and fasten with glue and 8d finishing nails. Cut the back from 1/4-inch leftover paneling or plywood. Square up the back with a carpenter's square and insert the back into the 1/4 × 1/4-inch rabbets cut in the sides and top and bottom. Glue and nail with 4d flat head nails.

Cut the storage top to size, making sure it fits properly into the back portion and extends out past the cabinet 1/2 inch on the sides and an extra 3/4 inch on the front for the facing that will be added. Round the top and bottom edges of the extended sides and front. Then fasten in place with glue and 8d finishing nails.

STEP 3
Adding the facing

Put the case on its back. Cut the pieces of facing to the given dimensions (7 pieces in all) and joint the edges with a plane or a jointer. With glue and 8d finishing nails, fasten the top front facer and the bottom front

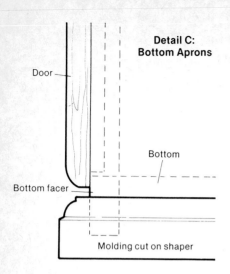

**Detail C:
Bottom Aprons**

Door

Bottom

Bottom facer

Molding cut on shaper

STEP 5
Completing the bookcase

Cut the doors to the given dimension (³⁄₈ inch wider and ³⁄₄ inch higher than the opening). Then cut a ³⁄₈-inch rabbet around the three sides of each door—leave as is the sides where the two doors join. Then, using a radius or quarter-round cutter in a shaper, round the outside edges of the doors. Install in place, using ³⁄₈-inch lip cabinet hinges fastened on the inside of the doors and to the outside of the cabinet facings. Bore holes for the pulls and install. Install magnetic catches on the inside of the cabinet behind to hold the doors closed. Cut the shelves to the correct size and round their front edges. Then sand and stain and finish to suit. To install the shelves in place, push in the shelf supports and drop the shelves down in place.

facer (top flush with the top edge, bottom flush with the top edge of the storage compartment). Use the same procedure to install the top side facing in position between the storage top and the top facer, with one edge of the facer flush with the edge of the side. Install the bottom door facer flush with the top of the bottom of the storage compartment. Finally, install the bottom side facers between the two horizontal facers on the storage compartment, edge flush with the side just as it is above.

STEP 4
Adding the aprons

Cut the two top side aprons to the given dimensions, cutting their front corners on a 45° miter, and use 6d finishing nails to nail and glue in place. Then cut the front and back top aprons to fit between them and glue and nail in place. Cut the stock for the bottom aprons about 3 inches longer than needed for each piece and shape their top edges using a glass bead cutter in a shaper, or leave them plain. Then cut the side aprons to the given length with a 45° miter on each outside front edge and glue and nail in place as shown in Detail C. Cut the front piece and glue and nail it in place. Set all nails about ¹⁄₈ inch below the wood surface with a fine nailset and fill in the holes, using a wood putty of the appropriate color. Then sand the entire case thoroughly.

BACKYARD STORAGE

Materials List

Base, 4″ × 6″ × 12′ (or double 2″ × 6″ × 12′ pressure-treated)
Floor nailers, 2″ × 2″ × 6′
Floor boards, 1″ × 4″ × 27′
Framing, 2″ × 6″ × 12′
Framing, 2″ × 2″ × 24′
Jamb, 1″ × 2″ × 6′
Wall braces, 2″ × 4″ × 16′
Wall sheathing, 1″ × 8″ × 128′
Roof nailer, 2″ × 2″ × 3′
† Roof nailer, 2″ × 4″ × 3′
† Fascia, 2″ × 6″ × 3′
Roof boards, 1″ × 8″ × 15′
Door frame, shelves, 1″ × 4″ × 28′
Door shelf lips, 1″ × 3″ × 8′
Door facing, 1″ × 8″ × 22′
Shelf (inside), 1″ × 8″ × 6″
Drawer sides, bottom, 1″ × 8″ × 14″
Drawer face, 1″ × 10″ × 2″
Drawer top, bottom frame, 1″ × 8″ × 18″
Drawer guides, 1″ × 2″
Front filler, 1″ × 6″ × 1″
Front filler, 1″ × 4″ × 1″
Cabinet top, bottom, shelf (1), 1″ × 8″ × 27″
Cabinet front, 1″ × 8″ × 6″
Cleats, stiles, 1″ × 2″ × 9″
3½″ butt hinges, with screws (1 pair), for doors
Galvanized steel strap for door hasp, ⅛″ × 1″ × 10″
No. 7 × ¾″ flat head wood screws for door hasp
Padlock for door (4)
2″ butt hinges, with screws (2 pairs), for cabinet doors (if needed)
Small hasp for cabinet doors (if needed)
Padlock for cabinet doors (if needed)
Metal or plastic flashing (6′)
Asphalt-impregnated felt roofing paper
Asphalt roofing cement
‡ Vinyl or metal downspout, elbow, drop outlet (10′)
* Splash block
6d, 8d galvanized common nails
6d, 8d galvanized finishing nails
⅜″ roofing nails

* For more than one module, front and back base pieces should be continuous.
† For more than one module, these pieces should be continuous.
‡ Only one needed for up to eight modules.

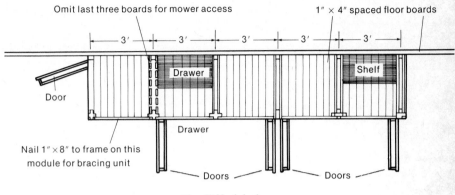

Omit last three boards for mower access

1″ × 4″ spaced floor boards

Drawer

Shelf

Door

Drawer

Nail 1″ × 8″ to frame on this module for bracing unit

Doors

Doors

Plan (5 Modules)

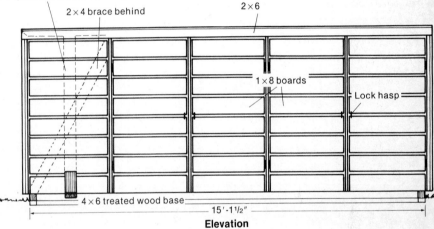

2″ × 4″ galvanized metal down spout

2 × 4 brace behind

2 × 6

1 × 8 boards

Lock hasp

4 × 6 treated wood base

15′-1½″

Elevation

Tools
Table saw, handsaw, or portable circular saw
Screwdriver
Hand sander

Length of Time Required
A few days

Safety Precautions
Safety goggles

Here is a very flexible plan to provide storage for all your garden tools, toys, supplies, furniture, and anything else that contributes to your backyard enjoyment. It is made up of 3-foot modules. The structure shown consists of five such modules, built against a horizontal-board fence—a perfect complement. But you can build more or fewer modules as need and space dictate. Of the units shown, one is fitted with a drawer (or it could be more), another with a cabinet, and the remainder with shelves both inside and on the doors—you can mix or match these elements as you please. And if you don't have a horizontal-board fence, you can anchor the modules to any other type of fence (although you may have to add some boards to cover the back), to a garage, or even to your house. As we said, very flexible. For that reason, the Materials List is given for a single module with one shelf, one drawer, and one cabinet. You take it from there.

STEP 1
Siting the module and laying the base

The structure should be built on level ground. Excavate to a level depth of 3 inches. (You may wish to excavate an area in front of the structure as well, as shown in the Side View. This area will be framed with pressure-treated 2×4s and filled to grade level with gravel, or it could be concreted.) Tamp the earth firm around the perimeter of the excavation. Lay in the pressure-treated 4×6 (or nail-laminated 2×6) pieces, toe nailing them together at the corners. Take care that all base pieces are level, scooping out earth under high points and filling in under low points with stones or broken-up concrete chunks. The base must be perfectly level with perfectly square corners before you proceed.

STEP 2
Assembling the module frame(s)

Assemble the wall framing as shown in the Top View and in the Side View. Rear framing members may be attached to fence posts or directly to the fence or other backing structure. Also attach the 2×4 rear roof nailer, making sure that it is level. The front uprights consist of 2×6s with 2×2s, 1×2s, or a combination of both (see

the Top View). Assemble these uprights before lifting into place and toe nailing to the base. Make sure that they are plumb (truly vertical) by checking with a carpenter's level in both directions (side-to-side, front-to-back). Nail the fascia across the front

uprights. Nail the wall sheathing boards to the framing.

STEP 3
Building the roof

Fasten the 2×2 roof nailer(s) inside the fascia, flush with the bottom.

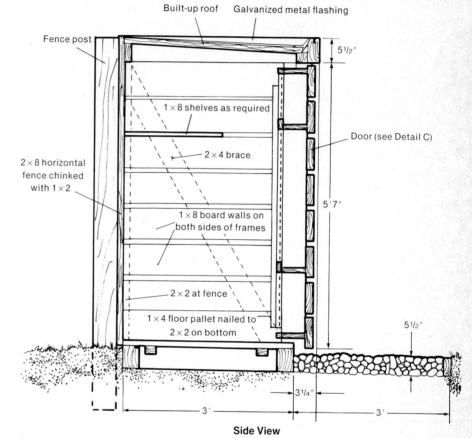

Side View

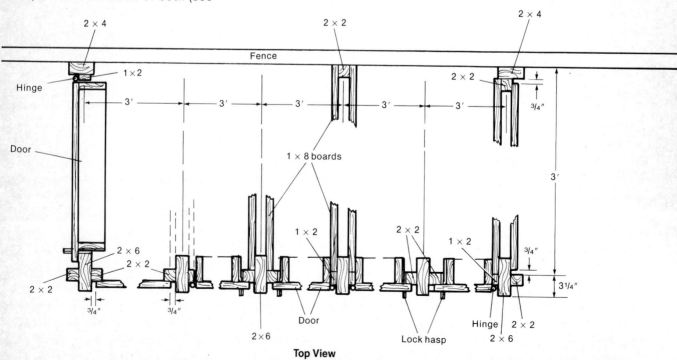

Top View

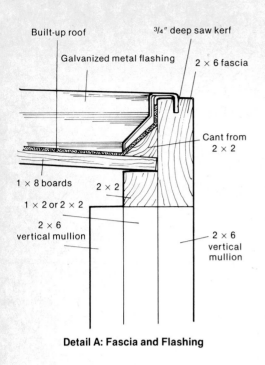

Detail A: Fascia and Flashing

Built-up roof

Galvanized metal flashing

¾" deep saw kerf

2 × 6 fascia

Cant from
2 × 2

1 × 8 boards

2 × 2

1 × 2 or 2 × 2

2 × 6
vertical mullion

2 × 6
vertical
mullion

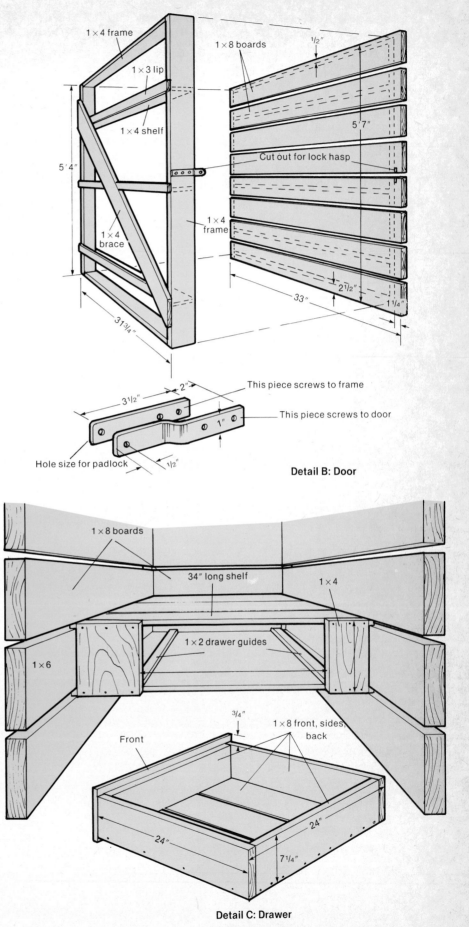

1 × 4 frame

1 × 3 lip

1 × 4 shelf

5′4″

1 × 4
brace

31¾″

1 × 8 boards

½″

5′7″

Cut out for lock hasp

1 × 4
frame

33″

2½″

1¼″

This piece screws to frame

This piece screws to door

3½″

2″

1″

Hole size for padlock

½″

Detail B: Door

1 × 8 boards

34″ long shelf

1 × 4

1 × 6

1 × 2 drawer guides

¾″

1 × 8 front, sides,
back

Front

24″

24″

7¼″

Detail C: Drawer

Nail roof boards in place. Install flashing at front and rear, as shown in the Side View and Detail A. Determine the best location for the downspout. In the structure shown, it is in the front wall of the far left module, which has a side-opening door. If all the drawers of your unit will open in the front, the downspout may be in a front corner. Cut a hole in the roof and insert the drop outlet. The downspout, elbow (through a hole cut in the siding), and splash block (to carry rain water runoff away from the structure) may be installed later. Nail a layer of roofing paper on the roof boards, cutting an opening around the drop outlet. Brush or mop on a coating of asphalt compound, then apply another layer of felt. Another coating of asphalt and a third layer of roofing are recommended.

STEP 4
Adding the floor

The floor is made up of individual pallets for each module, consisting of 1 × 4 boards nailed to 2 × 2s at front and rear (see the Side View). Before setting them in place, it is advisable to spread gravel over the earth in the excavation to discourage weed growth.

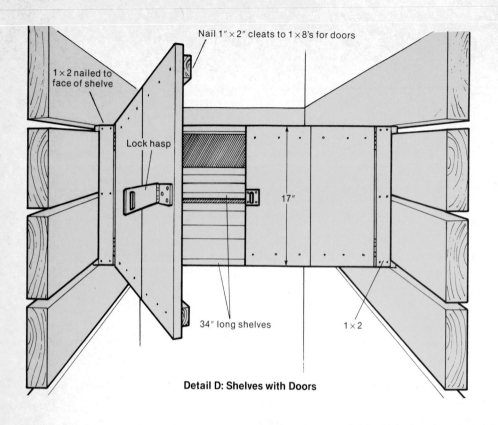

Nail 1″ × 2″ cleats to 1 × 8's for doors

1 × 2 nailed to face of shelve

Lock hasp

17″

34″ long shelves

1 × 2

Detail D: Shelves with Doors

STEP 5
Building the door(s)

Assemble the door frame with intermediate shelves as shown in Detail B. Make sure it is perfectly square, then nail on the diagonal brace. Add the shelf lips. Nail on the face boards. Hang each door with a pair of 3½-inch butt hinges. The dimensions of a specially crafted hasp are shown in Detail B. If you do not have any tools for bending metal, take it to a sheet metal shop; they can do the job in seconds on a brake. Fasten the hasp sections to the door and the frame.

STEP 6
Adding drawers and shelves

Assemble drawers as shown in Detail C, with a 1 × 10 front face overlapping at the bottom as a pull. Mount the drawers as shown in Detail C.

Detail D shows the cabinet construction. Doors are made of vertical 1 × 8s nailed to 1 × 2 cleats and mounted on 2-inch butt hinges to 1 × 2 stiles fastened to the shelves.

Shelves in the modules may simply be supported in the gaps between wall boards or mounted.

GLOSSARY

air-dried Lumber that has been dried by exposure to air, usually in the open, without artificial heat.

along-the-grain In the same direction as the grain; in plywood, the same direction as the grain of the face ply, usually the long dimension. Plywood is stronger and stiffer along the grain than it is across the grain.

annual growth ring The growth layer a tree puts on in a single year, including springwood and summerwood.

back The side of a plywood panel having the lower-grade veneer.

bark The outer layer of a tree, comprising the inner bark (the thin living part) and the outer bark or corky layer (composed of dead tissue).

bastard-sawn Hardwood lumber in which the annual rings are on an angle at or near 45° with the surface of the piece.

beam One of the principal horizontal structural support members of a building. *See also* **girder**.

bearing wall A wall that supports a floor or roof of a building.

bevel To cut edges or ends at an angle to make smooth mating joints between pieces.

bird's-eye A small localized spot in wood with the fibers indented and contorted around it to form circular or elliptical figures resembling the eye of a bird. Used for decorative purposes, these are common in sugar maple but rare in other hardwoods.

bleeding The seeping of resin or gum from lumber.

board A term generally applied to lumber 1 inch thick and 2 or more inches wide.

board foot Unit of lumber measurement: 1 foot × 1 foot × 1 inch.

bond To glue together, as veneers are "bonded" to form a sheet of plywood. Pressure can be applied during the process to keep mating parts in proper alignment.

bow The distortion in a board that deviates from flatness lengthwise but not across its faces.

bridging Wood or metal pieces placed diagonally between floor joists to hold them in line.

broad-leaved trees *See* **hardwoods**.

burl In wood or veneer, a severe localized distortion, generally rounded in form, of the grain, usually resulting from the entwined growth of a cluster of undeveloped buds. Such burls are the source of the highly figured burl veneers used for ornamental purposes.

butt joint A joint formed by abutting the squared ends or ends and faces of two pieces. Because of the inadequacy and variability in strength of butt joints when glued, they are not generally glued.

cambium The layer of tissue, one cell thick, between the bark and wood that repeatedly subdivides to form new wood and bark cells.

cell A general term for the minute units of wood structure, including wood fibers, vessel segments, and other elements of diverse structure and function.

chamfer The flat surface created by slicing off the square edge or corner of a piece of wood or plywood.

check A separation of the wood normally occurring lengthwise across the rings of annual growth; usually a result of seasoning.

clear lumber A term including the higher grades of lumber. It is sound and relatively free of blemishes.

closed-grained *See* **grain**.

coarse-grained *See* **grain**.

collar beam A horizontal beam fastened between pairs of rafters to add rigidity.

commons A term describing the ordinary grades of knotty lumber.

conifers *See* **softwoods**.

construction lumber Lumber that is suitable for ordinary and light construction.

continuous beam A beam supported at three or more points over two or more spans.

cripples Cut-off framing members above and below windows and in gables.

crook A distortion of a board in which there is a deviation edgewise from a straight line from end to end of the board.

crossband (cores) In plywood, the layers of veneer whose grain direction is at right angles to that of the face plies; they minimize shrinking and swelling.

cross-grained *See* **grain**.

cross lamination In plywood manufacture, the placing of consecutive layers at right angles to one another to minimize shrinkage and increase strength.

cup A curve in a board across the grain or width of a piece.

curly-grained *See* **grain**.

dado A joint formed by the intersection of two boards, in which one is notched with a rectangular groove to receive the other.

decay Disintegration due to the action of wood-destroying fungi; *dote* and *rot* are used to describe the same condition.

diagonal grain Annual rings at an angle with the axis of the piece, as a result of sawing at an angle with the bark of the tree.

dimension lumber A term generally applied to lumber 2 to 4 inches thick and 2 or more inches wide.

dressed lumber Lumber after shrinking from the green dimension and being surfaced with a planing machine, usually $3/8$ or $1/2$ inch smaller than the nominal (rough) size; for example, a 2 × 4 stud actually measures $1^{1}/_{2} \times 3^{1}/_{2}$ inches.

dry rot A term applied to many types of decay but especially to that which, when in an advanced stage, permits the wood to be easily crushed to a dry powder; the term is misleading because it is caused by moisture.

eased edges A term used to describe slight rounding of edge surfaces of a piece of lumber or plywood to remove sharp corners.

edge The narrow face of a rectangular piece of lumber.

edge-grained See **grain**.

encased knot A knot whose annual growth rings are not intergrown with those of the surrounding wood.

epoxy A synthetic resin used in some paints and adhesives because of its toughness, adhesion, and resistance to solvents.

face The wide surface of a piece of lumber; the side showing the better quality or appearance on which a piece is graded.

fascia Wood or plywood trim used along the eave or gable end of a structure.

figure The pattern produced in a wood surface by annual growth rings, rays, knots, deviations (such as interlocked and wavy grain), and irregular coloration.

fine-grained See **grain**.

finger joint A method of splicing pieces of lumber by machining the ends and gluing them together. The joint is similar to interlocking the fingers of two hands.

finish A term including the higher grades of sound, relatively unblemished lumber. Also, a material applied to protect and enhance the appearance of wood.

finished size The net dimensions after surfacing.

flat-grained See **grain**.

flitch A portion of a log, with or without bark, sawed on two or more sides and intended for remanufacture into lumber or veneer.

floor joists Framing members that rest on outer foundation walls and interior beams or girders.

frame construction A type of building in which the structural parts are wood or dependent on a wood framework for support. Commonly, lumber framing is sheathed with plywood or boards for floors, walls, and roof.

framing lumber The rough lumber of a house—joists, studs, rafters, and beams—usually 2 to 4 inches thick and at least 2 inches wide.

furring The process of leveling parts of a ceiling, wall, or floor by means of wood strips called **furring strips**, before adding finish material.

gable The triangular part of a wall under the inverted V of the roof line.

girder A large horizontal center beam used to support floor joists.

good-one-side Plywood that has a higher-grade veneer on the face than on the back; used where only one side will be visible. In identifying these panels, the face grade is given first.

grade The designation of the quality of lumber or plywood.

grain The direction, size, arrangement, appearance, or quality of the fibers in wood. To have a specific meaning the term must be qualified:

  closed-grained Wood with narrow, inconspicuous annual rings.

  coarse-grained Wood with wide annual rings in which there is considerable difference between springwood and summerwood.

  cross-grained Wood in which the fibers deviate from a line parallel to the sides of the piece in a diagonal or spiral grain or a combination of the two.

  curly-grained Wood in which the fibers are distorted so that they have a curled appearance, as in bird's-eye wood.

  edge-grained Lumber sawed parallel with the pith of the log and approximately at right angles to the growth rings so that the rings form an angle of 45° to 90° with the surface of the piece.

  fine-grained Another term for closed-grained wood.

  flat-grained Lumber sawed approximately perpendicular to the log's radius. A piece is considered flat-grained when the growth rings make an angle of less than 45° with the surface.

  open-grained Common classification for wood with large pores, such as oak, ash, chestnut, and walnut.

  straight-grained Wood in which the fibers run parallel to the axis of a piece.

  vertical-grained Another term for edge-grained lumber.

green lumber Lumber that has not been intentionally dried or seasoned, or that has been inadequately dried and tends to warp or bleed.

growth ring The layer grown on a tree in a single year, including springwood and summerwood.

hardwoods The botanical group of broad-leaved trees such as oak or maple.

The term has no reference to the actual hardness of the wood.

header A cross member placed between studs or joists for support over openings, as for a stairway, chimney, or door.

heartwood The nonactive core of a tree, usually darker and more decay-resistant than sapwood because gums and resins have seeped into it.

heel The part of a rafter that rests on the wall plate.

intergrown knot A knot whose rings of annual growth are completely intergrown with those of the surrounding wood.

joint The junction of two pieces of wood or veneer.

joist Horizontal framing member that supports a floor or the laths or panels of a ceiling.

joist spacing Distances (specified by building codes) between joists. Usually specified as **o.c.**

kerf A slot made by a saw; the width of the saw cut.

kiln-dried Wood that has been dried in ovens by controlled heat and humidity to specified limits of moisture content.

knot The portion of a branch or limb that has been surrounded by subsequent growth of the wood of the tree trunk.

lap To position two pieces so that the surface of one extends over that of the other.

lignin The second most abundant constituent of wood, located principally in the **middle lamella** (the thin cementing layer between the wood cells).

log A section of the trunk of a tree in suitable length for sawing into commercial lumber.

loose knot A knot that is not held firmly in place by growth or position and cannot be relied upon to remain in place.

loosened or raised grain A small section of the wood that has been loosened or raised, but not displaced.

lumber The product of the saw and planing mill not further manufactured than by sawing, resawing, passing lengthwise through a standard planing machine, crosscutting to length, and matching.

lumber-core Plywood construction in

which the core is composed of lumber strips and the outer plies are veneer.

millwork All building materials made of finished wood and manufactured in millwork plants or planing mills. This includes such items as inside and outside doors, window and door frames, mantels, stairways, and moldings. It does not include flooring, ceiling, or siding.

miter joint A joint formed by fitting together two pieces of lumber or plywood that have been cut off on an angle.

moisture content The amount of water contained in wood, usually expressed as a percentage of the weight of oven-dried wood.

molding A strip of decorative material with a planed or curved narrow surface for ornamental use. Moldings are often used to hide gaps at wall junctures.

on-center (o.c.) spacing The distance from the center of one structural member to the center of the adjacent member, as in the spacing of studs, joists, rafters, and so on.

open-grained *See* **grain**.

oven-dried wood Wood that has been dried so completely that it is without any moisture content.

panel faces Outer veneers of a plywood panel.

pin knot A knot that is not more than 1/2 inch in diameter.

pitch The angle of slope of a roof. Also, an accumulation of resin.

pitch pocket An opening extending parallel to the growth rings containing, or that has contained, either solid or liquid pitch.

pitch streak A localized accumulation of pitch in wood cells in a more or less regular streak.

pith The small, soft core in the structural center of a tree trunk, branch, or log.

plainsawed Lumber that has been sawed in a plane approximately perpendicular to a radius of the log (another term for **flat-grained**).

plank A broad board, usually more than 1 inch thick, laid with its wide dimension horizontal and used as a bearing surface (differs from joist in that latter is used on edge).

plate In wood frame construction, the horizontal dimension lumber member placed on the bottom and top of the wall studs to tie them together and to support the joists and rafters.

ply A single veneer in a glued plywood panel.

plywood A panel made of three or more layers of veneer joined with glue and usually laid with the grain of adjoining plies at right angles. To secure balanced construction, an odd number of plies is almost always used.

pole Round timber of any required size or length, used as a vertical support.

pore An indentation on the surface of a piece created by large-diameter, open-ended wood cells that form continuous tubes in a tree.

porous woods Woods that have vessels or pores large enough to be easily seen without magnification.

posts Large pieces (nominal dimensions: of 5 × 5 inches and larger, width not more than 2 inches greater than thickness) of square or approximately square cross section graded primarily for use as posts or columns.

preservative Any substance that, for a reasonable length of time, will prevent the action of wood-destroying fungi and borers, and other destructive insects when the wood has been properly coated or impregnated with it.

quartersawed Lumber that has been sawed so that the wide surfaces are roughly at right angles to the growth rings (another term for **edge-grained**).

rabbet A joint formed by cutting a groove in the surface along the edge of a board, plank, or panel to receive another piece.

radial In a round timber or piece of lumber, a line or surface extending outward from the core; a radial surface is always edge-grained.

rafter One of a series of structural roof members spanning from an exterior wall to a center ridge board.

raised grain A roughened condition of the surface of dressed lumber in which the hard summerwood is raised above the softer springwood but not torn loose from it.

rays Strips of cells extending radially within a tree and varying in height from a few cells to 4 or more inches.

resaw To reduce the thickness or width of boards, planks, or other material by cutting into two or more thinner pieces on a resaw.

ridge board The top horizontal board of a sloping roof to which the ends of the rafters are attached.

roof sheathing Boards or plywood nailed to the top edges of rafters to tie the roof together and support the roofing material.

rotary-cut veneer Veneer cut in a continuous strip by rotating a log against the edge of a knife in a lathe.

rough lumber Lumber as it comes from the sawmill; not further surfaced, machined, or dressed.

sap All of the fluids in a tree except special secretions and excretions such as gum.

sapwood The living wood of pale color near the outside of the log; generally more susceptible to decay than heartwood.

saw kerf Grooves or notches made when cutting with a saw. Also, the piece of wood removed by the saw in parting the material into two pieces.

sawed veneer Veneer produced by sawing.

scarf joint An end joint or splice formed by gluing together the ends of two pieces that have been tapered or beveled to form sloping plane surfaces, usually to a feather edge.

seasoning Removing the moisture from green wood in order to improve its serviceability.

second growth Usually, young trees that have grown after the removal by any means of all or a large portion of the previous stand.

select lumber The higher grades of sound, relatively unblemished lumber.

shake A separation along the grain, the greater part of which occurs between the growth rings.

sheathing The structural covering, usually boards or plywood, placed over exterior framing. Exterior facing material is attached to it.

shop lumber Lumber intended to be cut up for use in further manufacture.

sill plate The lowest member of the house framing resting on the top of the foundation wall (also called the **mud sill**).

sliced veneer Veneer that is sliced off

by moving a log or flitch against a large knife.

softness The property of wood that is indicated by a relative lack of resistance to cutting, denting, scratching, pressure, or wear.

softwoods The botanical group of trees that have needle or scalelike leaves. Except for cypress, larch, and tamarack, softwoods are evergreen. The term has no reference to the actual hardness of the wood.

solid-core Plywood composed of veneers over a lumber core.

sound knot A knot that is solid across its face, is at least as hard as the surrounding wood, and shows no indication of decay.

span The distance between supports.

spiked knot A knot cut approximately parallel to its long axis so that the exposed section is definitely elongated.

split A lengthwise separation of the wood, due to the tearing apart of the wood cells.

springwood The portion of the annual growth ring that is formed during the early part of the growing season and is weaker and less dense than summerwood.

straight-grained See **grain**.

structural lumber Cut chiefly from the middle portion of the log between the sap and the pith; yields high-strength construction lumber.

structural timber Pieces of wood of relatively large size, the strength of which is the controlling factor in their selection and use.

stud Generally, 2 × 4s used as the basic vertical framing members of walls. Studs are usually spaced either 16 inches or 24 inches o.c.

subfloor Boards or plywood sheets that are nailed directly to the floor joists and that serve as a base for the finish flooring.

summerwood The portion of the annual growth ring that is formed after springwood formation and generally is stronger and denser.

surfaced lumber Lumber that has been planed or sanded on one or more surfaces.

tight knot A knot so fixed by growth, shape, or position that it retains its place in the piece.

timber Lumber at least 5 inches in dimension; used in heavy construction.

toe nail Driving nails into corners or other joints at an angle.

tongue-and-groove A carpentry joint in which the jutting edge of one board fits into the grooved edge of a mating board.

twist A distortion caused by the turning or winding of the edges of a board so that the four corners of any face are no longer on the same plane.

veneer A thin layer or sheet of wood.

vertical-grained See **grain**.

virgin growth The original growth of mature trees.

wall sheathing Boards, plywood, or other material nailed to the outside face of studs as a base for exterior siding.

wane Bark or lack of wood from any cause on the edge or corner of a piece of wood.

warp Any variation from a true or plane surface. The term covers **crook**, **bow**, **cup**, **twist**, and any combination of these.

wavy-grain Wood in which the fibers collectively take the form of waves or undulations.

yard lumber Lumber of all grades, sizes, and patterns; intended for ordinary construction and general building purposes.

Picture Credits

INDEX

SALINAS PUBLIC LIBRARY
3 3550 00567 3939

SALINAS PUBLIC LIBRARY

LUMBER

Sizes: Metric cross-sections are so close to their nearest Imperial sizes, as noted below, that for most purposes they may be considered equivalents.
Lengths: Metric lengths are based on a 300mm module which is slightly shorter in length than an Imperial foot. It will therefore be important to check your requirements accurately to the nearest inch and consult the table below to find the metric length required.
Areas: The metric area is a square metre. Use the following conversion factors when converting from Imperial data: 100 sq. feet = 9.290 sq. metres.

METRIC SIZES SHOWN BESIDE NEAREST IMPERIAL EQUIVALENT

mm	Inches	mm	Inches
16 x 75	⅝ x 3	44 x 150	1¾ x 6
16 x 100	⅝ x 4	44 x 175	1¾ x 7
16 x 125	⅝ x 5	44 x 200	1¾ x 8
16 x 150	⅝ x 6	44 x 225	1¾ x 9
19 x 75	¾ x 3	44 x 250	1¾ x 10
19 x 100	¾ x 4	44 x 300	1¾ x 12
19 x 125	¾ x 5	50 x 75	2 x 3
19 x 150	¾ x 6	50 x 100	2 x 4
22 x 75	⅞ x 3	50 x 125	2 x 5
22 x 100	⅞ x 4	50 x 150	2 x 6
22 x 125	⅞ x 5	50 x 175	2 x 7
22 x 150	⅞ x 6	50 x 200	2 x 8
25 x 75	1 x 3	50 x 225	2 x 9
25 x 100	1 x 4	50 x 250	2 x 10
25 x 125	1 x 5	50 x 300	2 x 12
25 x 150	1 x 6	63 x 100	2½ x 4
25 x 175	1 x 7	63 x 125	2½ x 5
25 x 200	1 x 8	63 x 150	2½ x 6
25 x 225	1 x 9	63 x 175	2½ x 7
25 x 250	1 x 10	63 x 200	2½ x 8
25 x 300	1 x 12	63 x 225	2½ x 9
32 x 75	1¼ x 3	75 x 100	3 x 4
32 x 100	1¼ x 4	75 x 125	3 x 5
32 x 125	1¼ x 5	75 x 150	3 x 6
32 x 150	1¼ x 6	75 x 175	3 x 7
32 x 175	1¼ x 7	75 x 200	3 x 8
32 x 200	1¼ x 8	75 x 225	3 x 9
32 x 225	1¼ x 9	75 x 250	3 x 10
32 x 250	1¼ x 10	75 x 300	3 x 12
32 x 300	1¼ x 12	100 x 100	4 x 4
38 x 75	1½ x 3	100 x 150	4 x 6
38 x 100	1½ x 4	100 x 200	4 x 8
38 x 125	1½ x 5	100 x 250	4 x 10
38 x 150	1½ x 6	100 x 300	4 x 12
38 x 175	1½ x 7	150 x 150	6 x 6
38 x 200	1½ x 8	150 x 200	6 x 8
38 x 225	1½ x 9	150 x 300	6 x 12
44 x 75	1¾ x 3	200 x 200	8 x 8
44 x 100	1¾ x 4	250 x 250	10 x 10
44 x 125	1¾ x 5	300 x 300	12 x 12

METRIC LENGTHS

Lengths Metres	Equiv. Ft. & Inches
1.8m	5′ 10⅞″
2.1m	6′ 10⅝″
2.4m	7′ 10½″
2.7m	8′ 10¼″
3.0m	9′ 10⅛″
3.3m	10′ 9⅞″
3.6m	11′ 9¾″
3.9m	12′ 9½″
4.2m	13′ 9⅜″
4.5m	14′ 9⅓″
4.8m	15′ 9″
5.1m	16′ 8¾″
5.4m	17′ 8⅝″
5.7m	18′ 8⅜″
6.0m	19′ 8¼″
6.3m	20′ 8″
6.6m	21′ 7⅞″
6.9m	22′ 7⅝″
7.2m	23′ 7½″
7.5m	24′ 7¼″
7.8m	25′ 7⅛″

All the dimensions are based on 1 inch = 25 mm.

NOMINAL SIZE (This is what you order.) Inches	ACTUAL SIZE (This is what you get.) Inches
1 x 1	¾ x ¾
1 x 2	¾ x 1½
1 x 3	¾ x 2½
1 x 4	¾ x 3½
1 x 6	¾ x 5½
1 x 8	¾ x 7¼
1 x 10	¾ x 9¼
1 x 12	¾ x 11¼
2 x 2	1¾ x 1¾
2 x 3	1½ x 2½
2 x 4	1½ x 3½
2 x 6	1½ x 5½
2 x 8	1½ x 7¼
2 x 10	1½ x 9¼
2 x 12	1½ x 11¼

NAILS

NUMBER PER POUND OR KILO / **LENGTH AND DIAMETER IN INCHES AND CENTIMETERS**

Size	Weight Unit	Common	Casing	Box	Finishing	Size	Length Inches	Length Centimeters	Diameter Inches	Diameter Centimeters*
2d	Pound	876	1010	1010	1351	2d	1	2.5	.068	.17
	Kilo	1927	2222	2222	2972					
3d	Pound	586	635	635	807	3d	1/2	3.2	.102	.26
	Kilo	1289	1397	1397	1775					
4d	Pound	316	473	473	548	4d	1/4	3.8	.102	.26
	Kilo	695	1041	1041	1206					
5d	Pound	271	406	406	500	5d	1/6	4.4	.102	.26
	Kilo	596	893	893	1100					
6d	Pound	181	236	236	309	6d	2	5.1	.115	.29
	Kilo	398	591	519	680					
7d	Pound	161	210	210	238	7d	2/2	5.7	.115	.29
	Kilo	354	462	462	524					
8d	Pound	106	145	145	189	8d	2/4	6.4	.131	.33
	Kilo	233	319	319	416					
9d	Pound	96	132	132	172	9d	2/6	7.0	.131	.33
	Kilo	211	290	290	398					
10d	Pound	69	94	94	121	10d	3	7.6	.148	.38
	Kilo	152	207	207	266					
12d	Pound	64	88	88	113	12d	3/2	8.3	.148	.38
	Kilo	141	194	194	249					
16d	Pound	49	71	71	90	16d	3/4	8.9	.148	.38
	Kilo	108	156	156	198					
20d	Pound	31	52	52	62	20d	4	10.2	.203	.51
	Kilo	68	114	114	136					
30d	Pound	24	46	46		30d	4/4	11.4	.220	.58
	Kilo	53	101	101						
40d	Pound	18	35	35		40d	5	12.7	.238	.60
	Kilo	37	77	77						
50d	Pound	14				50d	5/4	14.0	.257	.66
	Kilo	31								
60d	Pound	11				60d	6	15.2	.277	.70
	Kilo	24								

*Exact conversion

WOOD SCREWS

SCREW GAUGE NO.	NOMINAL DIAMETER Inch	NOMINAL DIAMETER mm	LENGTH Inch	LENGTH mm
0	0.060	1.52	3/16	4.8
1	0.070	1.78	1/4	6.4
2	0.082	2.08	5/16	7.9
3	0.094	2.39	3/8	9.5
4	0.0108	2.74	7/16	11.1
5	0.122	3.10	1/2	12.7
6	0.136	3.45	5/8	15.9
7	0.150	3.81	3/4	19.1
8	0.164	4.17	7/8	22.2
9	0.178	4.52	1	25.4
10	0.192	4.88	1 1/4	31.8
12	0.220	5.59	1 1/2	38.1
14	0.248	6.30	1 3/4	44.5
16	0.276	7.01	2	50.8
18	0.304	7.72	2 1/4	57.2
20	0.332	8.43	2 1/2	63.5
24	0.388	9.86	2 3/4	69.9
28	0.444	11.28	3	76.2
32	0.5	12.7	3 1/4	82.6
			3 1/2	88.9
			4	101.6
			4 1/2	114.3
			5	127.0
			6	152.4

Dimensions taken from BS1210; metric conversions are approximate.

BRICKS AND BLOCKS

Bricks

Standard metric brick measures 215 mm x 65 mm x 112.5. Metric brick can be used with older, standard brick by increasing the mortaring in the joints. The sizes are substantially the same, the metric brick being slightly smaller (3.6 mm less in length, 1.8 mm in width, and 1.2 mm in depth).

Concrete Block

Standard sizes

390 x 90 mm
390 x 190 mm
440 x 190 mm
440 x 215 mm
440 x 290 mm

Repair block for replacement of block in old installations is available in these sizes:
448 x 219 (including mortar joints)
397 x 194 (including mortar joints)

PIPE FITTINGS

Only fittings for use with copper pipe are affected by metrication: metric compression fittings are interchangeable with Imperial in some sizes, but require adaptors in others.

INTERCHANGEABLE SIZES mm	INTERCHANGEABLE SIZES Inches	SIZES REQUIRING ADAPTORS mm	SIZES REQUIRING ADAPTORS Inches
12	⅜	22	¾
15	½	35	1¼
28	1	42	1½
54	2		

Metric capillary (soldered) fittings are not directly interchangeable with imperial sizes but adaptors are available. Pipe fittings which use screwed threads to make the joint remain unchanged. The British Standard Pipe (BSP) thread form has now been accepted internationally and its dimensions will not physically change. These screwed fittings are commonly used for joining iron or steel pipes, for connections on taps, basin and bath waste outlets and on boilers, radiators, pumps etc. Fittings for use with lead pipe are joined by soldering and for this purpose the metric and inch sizes are interchangeable.
(Information courtesy Metrication Board, Millbank Tower, Millbank, London SW1P 4QU)

MAY 1993
Received
John Steinbeck
Library